Beyond Flagpoles and Footprints

Pioneering the Space Frontier

Buzz Aldrin and Larry Bell

Beyond Flagpoles and Footprints: Pioneering the Space Frontier

Also by Larry Bell

Scared Witless: Prophets and Profits of Climate Doom
Cosmic Musings: Contemplating Life beyond Self
Climate of Corruption: Politics and Power behind the Global Warming Hoax
Reflections on Oceans and Puddles: One Hundred Reasons to be Enthusiastic, Grateful and Hopeful
Reinventing Ourselves: How Technology Is Rapidly and Radically Transforming Humanity
What Makes Humans Truly Exceptional: Including Us
Cyberwarfare: Targeting America, Our Infrastructure and Our Future
Thinking Whole: Rejecting Half-Witted Left & Right Brain Limitations
The Weaponization of AI and the Internet: How Global Networks of Infotech Overlords Are Expanding Their Control Over Our Lives

Also by Buzz Aldrin

No Dream Is Too High: Life Lessons From a Man Who Walked on the Moon (with Ken Abraham)
Reaching for the Moon (illustrated by Wendell Minor)
Mission to Mars: My Vision for Space Exploration (with Leonard David)
Welcome to Mars: Making a Home on the Red Planet (with Marianne Dyson)
Magnificent Desolation: The Long Journey Home from the Moon (with Ken Abraham)
Return to Earth (with Wayne Warga)
Look to the Stars (illustrated by Wendell Minor)
Encounter with Tiber (with John Barnes)
Men from Earth
The Return (with John Barnes)

© 2021 Buzz Aldrin and Larry Bell All Rights Reserved
Print ISBN 978-1-941071-55-7
eBook ISBN 978-1-941071-56-4

Cover illustration by Albert Rajkumar
Cover Design by Guy D. Corp
www.GrafixCorp.com

STAIRWAY≡PRESS

www.StairwayPress.com
1000 West Apache Trail
Suite 126
Apache Junction, AZ 85120

Buzz and Larry with Neil Armstrong's statue at Purdue Unversity

Buzz and Larry

BUZZ ALDRIN, GLOBALLY renowned as one of the first two humans to travel to the Moon, is a West Point graduate, former USAF Korean War combat pilot, MIT doctorate recipient and NASA Gemini 12 mission pilot.

Buzz undertook three pioneering Gemini spacewalks, was instrumental in establishing underwater astronaut training, developed the *Aldrin Cycler* Mars spacecraft trajectory concept and is Presidential Medal of Freedom recipient.

Larry Bell is an Endowed Professor of Space Architecture at the University of Houston where he founded the Sasakawa International Center for Space Architecture and Space Architecture Graduate Program.

Larry, former NASA Johnson Space Center Chief Engineer Max Faget, and two other partners also co-founded Space Industries, Inc., where Neil Armstrong and two retired NASA JSC Directors served as board members.

Special Acknowledgements

THIS BOOK HAS been a long time in the making and is the product of a close friendship full of endless conversations dating back more than four decades.

I suggested the idea of writing this history and commentary to Buzz in one of those discussions a few years ago, addressing how global space developments that have profoundly influenced our worst fears and highest hopes originated…and where they might lead.

In this broad context, international space developments both drive and mirror human society, providing both lethal weapons of destruction and limitless wonders of discovery.

Numerous books and articles have been written about Buzz Aldrin, and very deservedly so.

I wish to call particular attention to an excellent one titled *Buzz Aldrin: Mission to Mars* authored by Leonard David and published by National Geographic (2013).

Several quotations of *Buzzword* statements which appear in this book are appreciatively taken—and are appropriately cited from that source—with Buzz's permission.

Most particularly, I appreciate some observations offered by Andrew Aldrin about his father in the foreword of Leonard's book which exactly capture a special aspect of the man I and others who know him well so deeply admire.

Andy wrote:

> *Over the 40 years of conversations about space, I really can't remember a single time that dad talked to me about his trip to the Moon. Sure, there were brief words here and there, but the conversation was always about the future. He cares where we are going as a civilization, not where we have been.*

Andy adds:

> *His vision is not just about the technical or programmatic elements; it is about the political and social underpinnings necessary to reinvigorate the nation's commitment to the human exploration and development of space.*

I recently mentioned to Buzz that his vision, in my opinion, may well prove to be an equal or even more enduring achievement legacy than his valor in those historic footprints that he—along with another friend, Neil Armstrong—first left on the Moon.

This reference was directed to Buzz's remarkably visionary concept of "Aldrin Cyclers," spacecraft guided by elegant and immutable laws of Newtonian physics perpetually ferrying crews and cargo back and forth between Earth, Moon and Mars along future space superhighways that neither he, nor I will live to see.

In retrospect, reflecting upon my own lessons learned in writing this book, there are a few, among many foundational technical visions and visionaries that warrant specially highlighted recognition.

Included, are: a 19th century small village Russian mathematics school teacher named Konstantin Tsiolkovsky who envisioned "rocket trains" (the concept of stageable rockets even before rocketry existed) that led to virtually everything that followed; American "father of the liquid-fueled rocket" Robert Goddard, who proved that spaceflight could be more than a dream; the NASA Apollo program's great engineer, John Houbolt, who came up with the bold idea of "lunar orbit rendezvous" that applied Tsiolkovsky's staged rocketry to deliver American astronauts to the Moon and back; private entrepreneur Elon Musk who demonstrated that rocket first stages can be recycled to make spaceflight more resourcefully sustainable; and yes, those *Aldrin Cyclers* that offer promises of rocket trains operating on endless round-trip orbital tracks that lead to future destinations of human exploration and evolution that our children and theirs will one day pioneer.

As a personal acknowledgement, I wish to especially thank Albert Rajkumar, a graduate of our University of Houston's Sasakawa International Center for Space Architecture (SICSA) program who provided much valued research contributions to incorporate a flurry of space development updates to help keep this manuscript as current as possible prior to release, and also for creating the cover graphic for this and some other of my books.

In addition, I am grateful for essential and continuing support from Ken Coffman, the Stairway Press publisher, for my most recent ten books.

—Larry Bell

The Visionaries		
1880s	▶	Sokolsky conceptualizes liquid fueled rockets.
1903	▶	Tsiolkovsky calculates possibility of orbital launch.
1926	▶	Goddard Launches first liquid fueled rocket.

The Space Race		
1936	▶	Germany starts the V-2 program.
1945	▶	Relocation of German Scientists post WW2.
1957	▶	Korolev and co. launch Sputnik-1 into orbit.
1958	▶	Creation of NASA.
1959	▶	Luna-1 becomes first spacecraft to orbit the Sun (when it misses the Moon).
1959	▶	Luna-2 becomes first spacecraft to make contact with Moon or any other celestial body.
1961	▶	Gagarin becomes first man in space.
1962	▶	Mars-1 makes Mars fly-by.
1962	▶	John Glenn becomes first American to orbit Earth. 1962.
1964	▶	Alexy Leonov becomes first person to perform spacewalk.
1964	▶	Zond-1 reaches vicinity of Venus.
1968	▶	Apollo 8 orbits around the Moon with humans.
1969	▶	Apollo 11 delivers the first humans on the Moon.
1971	▶	Salyut-1 becomes world's first space station.
1973	▶	Skylab becomes first American space station.

International Competition and Cooperation		
1975	▶	Apollo-Soyuz rendezvous.
1981	▶	First Space Shuttle mission. 1981
1983	▶	Mir Space Station launched. 1986.
Late 1980s	▶	NASA's commercialization of space precedence set by Space Industries, Inc. Late 1980's.
1982	▶	SSIA becomes first private enterprise to have a payload in orbit.
1986	▶	Space Shuttle Challenger disaster.
1998	▶	Buran flies.
1998	▶	International Space Station begins life in space.
2000	▶	MirCorp becomes first private venture to fund a manned expedition to a space station; a resupply mission in space; a spacewalk; and the first space tourist.
2000s	▶	Emergence of China in Space.

	2002	Founding of SpaceX
	2004	Virgin Galactic becomes first privately funded company to make a sub-orbital flight.
	2005	NASA announces Constellation program for sustainable space exploration.
	2008	NASA competitively awards contracts for private ISS cargo services through CRS program.
	2008	SpaceX becomes first privately funded company to launch a rocket into orbit.
Commercialization	2010	SpaceX becomes first privately funded company to successfully orbit and recover a spacecraft.
	2012	SpaceX becomes first privately funded company to send a spacecraft to the ISS.
	2012	SpaceX soft-lands Grasshopper, a Vertical Take off Vertical Landing (VTVL) prototype for Starship.
	2013	SpaceX becomes first privately funded company to launch a satellite into LEO.
	2015	SpaceX safely lands and recovers a Falcon 9 first stage, ushering in age of reusable rockets.
	2016	Bigelow Aerospace first docks an inflatable module to the ISS.
	2019	SpaceX successfully completes StarHopper test flight to demonstrate Starship development.
	2021	SpaceX becomes first private company to deliver humans to ISS.
	2021	Virgin Galactic becomes first private company to fly space tourists to space in a sub-orbital flight.
	2021	Blue Origin flies, among others, its CEO, Jeff Bezoz, and on a later flight, William Shatner who played Captain Kirk in Star Trek on sub-orbital space trips.

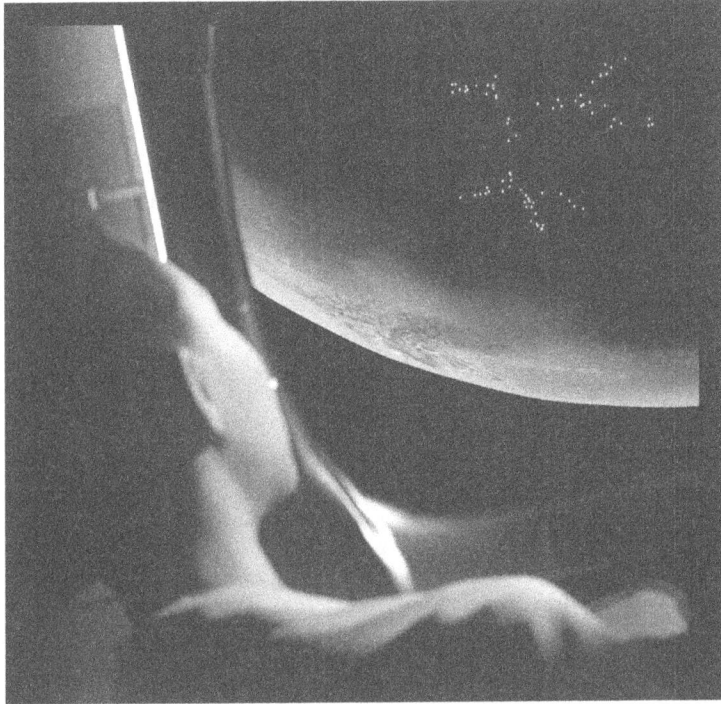

Introduction

CONCEIVED IN THE minds of visionaries and gestated in wombs of governments during times of international conflict, humankind gave birth to a cosmic dream child of peace that would mature to achieve inspirational goals, influencing the entire world in previously unimaginable and unquestionably beneficial ways. Orbiting satellites have erased global communication boundaries; spawned a transformative internet information-sharing network; monitored natural and man-made events that affect our safety; coordinated and guided air and surface transportation movements; and have supported unlimited international business opportunities.

Advancements in rocketry, spacecraft and instruments of exploration have opened an epic new era of cosmic discovery. And yes, the complex challenges driving such achievements have indeed yielded countless technological advancements and business opportunities that continue to enhance the quality of our everyday lives.

In total, these advancements have expanded human experience while making our world seem smaller.

Such developments came as unforeseen results arising from a far different motivation, one prompted by events originating in the former USSR that harshly jolted the American psyche. The shock waves began on October 4, 1957 when a tiny Soviet satellite chirped alarming evidence of technological superiority.

Soon afterwards, an emboldened Soviet President Nikita Khrushchev banged that point home with his shoe on a table at a 1960 UN meeting, crowing *We will bury you.* Then, only one year later, a young cosmonaut named Yuri Gagarin really rubbed it in, opening a new extraterrestrial era that threatened to leave the U.S. behind.

America responded to the challenge.

On May 25, 1961, only a few weeks after Gagarin's flight, President John Kennedy upped the ante, committing the U.S. to send a man to the Moon and returning him safely before the end of that decade. We did even better…putting four of our citizens on the lunar surface and returning them by 1969, along with two others in lunar orbit and back on the same Apollo missions. Within three more years, eight others had walked on the Moon on successful round-trip voyages with four more orbital companions.

April 1970 witnessed remarkable flight and ground crew resilience and ingenuity during an aborted Apollo 13 mission. In preparation, four Earth-orbital Mercury launches (carrying one astronaut each), one suborbital and nine Earth-orbital Gemini flights (two astronauts each), two Earth-orbital Apollo tests (three astronauts each), and two orbital tests made the lunar landings possible.

That was only the beginning.

The Skylab program (1973-79) established America's first true space station and demonstrated human abilities to adapt and undertake productive work under orbital weightless conditions. Apollo-Soyuz enabled U.S. and Russian engineers and flight crews to rise above Cold War rivalry and work together in a literal high ground. American astronaut visitations to the Russian MIR space station extended this spirit of cooperation and diplomacy. Development and assembly of the International Space Station (ISS) now culminates the largest, most complex initiative in human history, a testament to great potentials and peaceful benefits of multi-national collaboration.

Yes, space exploration programs have produced technological innovations, but even more, they have inspired our children and grandchildren, and can do so for future generations to realize that the sky is literally no limit to what can be achieved with ambitious goals, solid educational foundations, and disciplined and sustained commitment. They will be the ones who advance future innovation and progress in all fields…the ones who will provide the vision, leadership and innovation which will determine our nation's destiny.

Then there's the matter of national security and prestige reflected by technological superiority. After all, that priority really got our space program off the ground from its inception.

Here, space exploration has served to stimulate interests of young people in science and engineering-based studies, providing lessons and problem-solving challenges that apply at all levels of learning. The impact has resulted in U.S. universities having well deserved international reputations. Less fortunately, many of the technical programs in these top-ranked institutions are now dominated by a majority of students from Asia and India.

As we Americans recognize symbolic and real benefits of international cooperation and national prestige gained through space exploration leadership, the future of such programs will determine something far more important than the way the rest of the world views us.

It will ultimately influence how we see ourselves.

Table of Contents

Part One: The Visionaries

THIS BOOK CHRONICLES an epic, unfinished story of remarkable people who have and will continue to charter new courses of progress and destiny.

Behind and among the experiences and examples of those very few referenced in these pages stand countless others whose contributions of purpose, passion, professionalism, and persistence have brought humankind to our present crossroads of great possibilities and uncertainties.

Prominently and appropriately featured in connection with early developments are three international leaders, Russia's Chief Designer Sergei Pavlovich Korolev, the Father of American Space Flight Robert H. Goddard, and head German V-2 rocket developer Wernher von Braun who independently converted brilliantly innovative concepts of a solitary Soviet school teacher into a technological reality which has inexorably changed and expanded our world.

This is also a story about historic achievements, events and lessons born of triumphs and tragedies. It reveals a nexus of politically manipulated and ideologically shifting public rivalries between nationalistic pride and paranoia where space exploration and technology manifests full dimensions of civilizations' boldest dreams and greatest fears.

As historian Walter A McDougal has observed in his 1985 book *Heavens and the Earth*, there is probably no more exemplary and ironic time and place to begin this narrative saga than in early Bolshevik Russia:

> *Modern rocketry and social revolution grew up together in tsarist Russia. There is no anomaly in the fact that the most 'backward' of the Great Powers before World War I was the one that fostered violent rebellion against the chains of human authority and the chains of nature.*

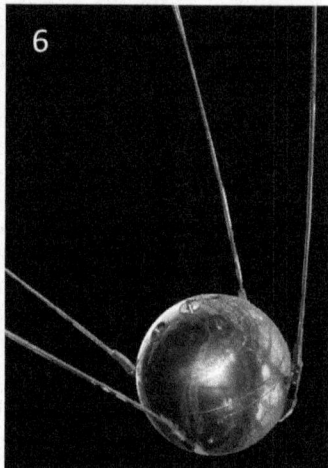

1. Sergei Korelev
   Source: Former USSR
2. Robber Goddard
   Source: NASA
3. Wernher von Braun
   Source: NASA
4. Konstatin Tsiolskovsky
   Source: NASA
5. Alan Shepard
   Source: NASA
6. Sputnik-1
   Source: NASA
7. Mir Space Station
   Source: NASA
8. V-2 Rocket
   Source: NASA

Chapter One: Tsiolkovsky's Remarkable Vision

WHEREAS RUSSIAN THINKERS dating back to the 1880s including Viktor Sokolsky contemplated general possibilities of creating liquid-fueled rockets which are commonplace today, we can thank the writings of a self-taught high school mathematics and physics teacher in the small town of Kaluga south of Moscow for the concepts and calculations upon which such realities depend.

Broadly considered to be the Father of Space Travel, Konstantin Eduardovich Tsiolkovsky (1857-1935) originally preoccupied his early years with personal design studies related to research into stellar radiation and design concepts for steam engines and metal-fabricated dirigibles. Then, upon conceptualizing possibilities for reaction-driven devices later called rockets, he published an amazing book in 1883 titled *Free Space* which proposed the design for a liquid-fueled propulsion device for use in the vacuum of space…

> …*which, when combined chemically would yield per unit mass of resultant product such an enormous amount of energy.*[1]

Tsiolkvosky's concept was comprehensive, detailed, and ingenious, comprising:

> …*an elongated configuration of least aerodynamic drag; liquid hydrogen and liquid oxygen propellant supplied to the combustion chamber by mixer valves; divergent nozzle; [and even providing a] cosmonaut's compartment.…*

Konstantin Tsiolkovsky's prolific productivity during the 1920s through early 30s was amazing, conceiving ideas for *a reaction engine* (1927-28); *a new airplane* (1928); *a jet-propelled aeroplane* (1929); *the theory of the jet engine* (1930-34); *the maximum speed of a rocket* (1931-33); and the final classic work before his death, *space rocket trains* (1924-1934).

He told a group of students at the Zhukovsky Academy in 1934:

> …*I am not at all sure, of course, that my 'space rocket train' will be appreciated and accepted readily, at this time. For it is a new conception reaching far beyond the present ability of man to make such things. However, time ripens everything; therefore, I am hopeful that some of you will see a space train in action.*[2]

And they did.

Tsilkovsky's rich legacy of design contributions guided design practices of a great officially unnamed Soviet engineer who led efforts which produced the USSR's first ICBM and launched the Space Age with the first orbiting satellite (Sputnik), the first dog, first man, first two men, first woman, first three men, first spy and communication satellites, and the vehicles and spacecraft that first reached the Moon and Venus and passed by Mars.

As noted by Chief Designer Sergei Pavlovich Korolev who was known secretly as "SP" by his team: Tsiolkovsky's principles included a rocket nozzle in...

...a special explosion tube in the form of a flared cone...

and...

...a combustion chamber in which fuel was supplied by pumps, suggested automated control of combustion in the engine, to gear its operation in the different conditions encountered during the rocket's flight along its trajectory.[3]

Further crediting Tsiolkovsky, Korolev applied "grids with slanted vents to be installed at the entrance of the explosion tube to create the most favorable conditions for combustion," plus the concept of regenerative cooling of the rocket by circulating cold fuel around the hot combustion chamber. Tsiolkovsky's visionary writings even suggested possible use of atomic energy (solar radiation and energy from space) to propel rockets; gyroscopic stabilization of rocket ships; control of flight by insertion of tilting surfaces in the rocket exhaust.

And that wasn't nearly all. In Korolev's descriptions, Tsiolkovsky directed amazingly comprehensive attention to entire spaceships and the well-being of human occupants. Crew safety accommodations included provisions for centrifuges to test effects of gravity on living organisms; protection for space travelers from the effect of rapid acceleration during launch and creation of an inside artificial gravity field to counter effects of prolonged weightlessness.

Tsiolkovsky providently conceived *space rocket trains* are now the standard multi-staging technique used to deliver payload elements to Earth orbits and planets. Included were *interplanetary stations* that would be:

...a combination of several rockets linked together after being placed in orbit...The station's orbit could be changed by burning an additional small quantity of fuel...communications between the station and Earth could be maintained by small rockets.[4]

Chapter Two: Korolev: Russia's Secret Hero

KONSTANTIN TSIOLKOVSKY'S GREAT conceptual achievements led to the USSR's first Salyut orbital space station which was realized in 1971 during the Chief Engineer Korolev's last years of life. Yet unlike Korolev, he did live to publicly receive high honors: appointment to the Soviet Academy of Sciences in 1919 and the Kremlin's coveted Order of the Red Banner of Labor award in 1932. Korolev's identity was concealed as a state security secret under orders from Stalin, Khrushechev and Brezhnev until Korolev's untimely death and burial at the Kremlin Wall. This occurred in 1966, at age 59, during the peak period of a USSR race to beat America in landing its citizens on the Moon. Only then was he publicly honored as Hero of Socialist Labour.

Korolev's early death might be attributed in part to health problems arising from brutal imprisonment conditions suffered as a victim of an oppressive Stalin regime. This is particularly ironic given his important prior and subsequent contributions to advance his country's military aviation and ballistic missile programs.

Sergei Pavlovich Korolev's remarkable career began with a childhood passion for aviation. As a young man he had hoped to enroll at the Zhukovsky Academy in Moscow where Andrei Tupolev and other experts served as professors while also designing military aircraft. However, lacking typical military pilot credentials held by other applicants, he enrolled instead at the Kiev Polytechnic Institute which, while less impressive, had an aviation section where a professor named N.B. Delone had built a biplane glider in 1909. That was only one year after the Wright Brothers demonstrated their airplane in Paris.[5]

In 1926, young Korolev transferred to the Moscow Higher Technical School, which offered a program in theoretical aeronautics and one of the first USSR wind tunnels. In 1927, one year after America's Robert Goddard successfully launched the world's first liquid-fueled rocket, Korolev attended an inspirational lecture by inventor Gyorgi Polevoi and the inventor of a rocket-powered automobile, Alexander Fedorov, titled *From Human Flight in the Air to Flights in Universal Ether*. Korolev's main interest envisioned rockets as means to improve aircraft performance. By 1929 he developed his own unpowered glider.

The 1930s witnessed the first steps toward developing liquid rocket engines in the USSR led by amateurs. Korolev joined a Group for Studying Rocket Propulsion (GIRD) and began working on a small gasoline-fueled GIRD-09 engine designed by Mikhail Tikhonravov weighing 40 pounds (18 kg) which flew for 18 seconds. On August 25, 1933, eight days after that historic first USSR launch in the Nakhabino woods, Korolev wrote an article titled *Towards the Rocketplane* which presented highly optimistic future prospects. He predicted:

> *Jet flight vehicles can develop flight speeds of 3,600 km/hr...and [can attain] immense altitudes [but that] practical resolution of this huge problem requires years of persistent work.*[6]

The efforts by Korolev and his coworkers at GIRD drew the attention of Russian military Marshal Mikhail Tukhachevsky (1893-1937) who began to fund follow-up developments of the 09 rocket aimed at achieving a series of launches with flight speeds of 1800-2000 mph (800-900 m/s) and hundreds to thousand of miles range. To help accomplish this, GIRD was merged into a new organization called the Reaction Propulsion Institute (RNII) headed by Ivan Kleimenov with Korolev serving as his Deputy Chief Engineer. Much of this activity was directed to creating reliable and accurate guidance and control systems for military applications.

Progress at RNII would later languish for seven years due to tragic anti-intelligencia circumstances which impacted thousands of USSR scientists, engineers and military leaders who fell under Stalin's purge...very much including RNII's sponsor Marshal Tukhachevsky—broadly considered to be one of the most heroic and intelligent officers in the USSR military—was executed on trumped-up charges of spying for the Germans. RNII's Deputy Chief Engineer Korolev was fortunate to survive brutal incarceration.

On the early morning of June 27, 1938, two KGB agents and two "witnesses" arrested 31-year-old Korolev without time to say good-by to his three-year-old daughter Natasha. He was accused of collaborating with an anti-Soviet organization in Germany in order to subvert a new field of technology. This event may not have been entirely unexpected following the arrest and prison sentencing three months earlier of Valentin Glushko, another leading Soviet rocket designer.[7]

As with Glushko and others charged with such offenses, Korolev received no trial and was beaten and forced to confess. After receiving a 10-year sentence and having his family property confiscated, he was moved from one prison to another. In October 1939, he was transferred to the most dreaded of all, the Kolyma forced labor camp in far eastern Siberia made infamous in the West in Aleksandr Solzhenitsyn's publication of *The Gulag Archipelago.*

Stalin's biographer Dmitri Volkogonov estimated that in order to evoke terror and obedience, around 4.5 to 5.5 million arrests occurred between the years 1937-1938 alone. Between 8- and 9-thousand of those charged received death sentences. Among others held captive, about 10 percent perished annually from malnutrition, execution, brutal discipline, along with inadequate food, shelter and clothing. Several thousand prisoners at Kolyma reportedly died each month...as many as 30 percent per year.[8]

Korolev's permanent health damage—traceable to his five-month cold winter ordeal cutting trees, digging and pushing wheelbarrows at a Kolyma gold mine—included a heart condition, broken jaw and loss of all teeth. Yet he was never known to speak to anyone about his hard treatment and privations until later, just a few days following a fifty-ninth birthday party. Late that night after other guests had departed, he confided a sad account to the world's first Earth-orbiting human, Yuri Gagarin.

After his appointment to a high RNII position Korolev had been blamed for spending too much money and was taken to Lefortovo prison where he was interrogated and beaten. Upon asking for a glass of water he was hit on the head by a jug handed to him and called an enemy of the people. He was then told "Today is your trial" and was led down a long corridor into a room. When the door opened, Kliment Voroshilov, one of Stalin's closest associates entered. Imagining that Voroshilov would straighten out the problem Korolev told him "I didn't commit any crime." Voroshilov then shouted "None of you swine [svolochi in Russian] have committed a crime. Ten years hard labour. Go! Next!"

Following physically and emotionally grueling Kolyma experiences and numerous failed appeals, Korolev's case was reinvestigated, and his sentence was reduced from 10 to 8 years in 1939 when Lavrentiy Beria was replaced by Nikolai Yezhov as Minister of Internal Affairs. He was then moved to greatly improved conditions at a penal institution known as Central Design Bureau 29 where most occupants were intellectuals, including scientists and engineers, and put to work in charge of wing design for a light bomber.

Korolev was later relocated to another penal institution in Kazan, Siberia when Germans approached Moscow. Although technically freed in 1942, he voluntarily elected to stay in order to continue work he considered important. On January 8, 1943 he was elevated to become chief designer for a Group No. 5 which worked on aircraft engines. Then in 1944 he was moved once again to a penal facility converted from the former hunting lodge of the last Russian emperor, Nicholas II, located at Krasnaya Polyana in the Caucasus, where he at once again began working as a rocket engineer until his formal release as a prisoner.

The following year, Korolev, who had previously been accused of collaborating with anti-Soviet Germans, was commissioned as a colonel in the Red Army and flown to Germany to gather information on von Braun's V-2 rocket developments.

In reality, KGB agents had been monitoring these activities since the 1930s.

Chapter Three: Goddard—America's Father of Spaceflight

GERMAN AND RUSSIAN liquid-fueled rocket development began years after Robert Goddard launched the world's first one on March 16, 1926 from his aunt Effie's farm in Auburn, Massachusetts. This was five years before Johannes Winkler launched Germany's first at Dessau in 1931, and seven years before the Soviet Union's GIRD-09 1933 success in the Nakhabino woods.

As Goddard described his launch to sponsor Charles G. Abbot at the Smithsonian Institution:

> *After about 20 seconds the rocket rose without perceptible jar, with no smoke and with no apparent increase in the rather small flame, increased rapidly in speed, and after describing a semicircle, landed 184 feet from the starting point—the curved path being due to the fact that the nozzle had burned through unevenly, and one side was longer than the other. The average speed, from the time of flight measured by a stopwatch, was 60 miles per hour. This test was very significant, as it was the first time a rocket operated by liquid propellants traveled under its own power.*[9]

All of these were extremely tiny and crude devices by today's standards. For comparison, Goddard's rocket weighed only 16 pounds (7kg)—including 10.25 pounds (4.65 kg) of liquid oxygen and gasoline—flew for 20 seconds, while the Soviet GIRD-09 gasoline and liquid oxygen-fueled rocket weighed 40 pounds (18 kg) and flew for 18 seconds.

Nevertheless, paraphrasing the immortal words of Neil Armstrong upon reaching the Moon's surface on July 20, 1969, those small steps indeed led to giant leaps. And as young Sergei Korolev predicted following GIRD's Nakhabino achievement, that persistent work required for jet flight vehicles to attain truly high flight speeds and altitudes has really paid off.

Some financial support from the Guggenheim family—thanks to an endorsement from American aviation hero Charles Lindberg—enabled Robert Goddard to leave his professorship position at Clark University in Worcester, Massachusetts and move his rocket development work to Roswell, New Mexico in 1931.

However, lack of success in obtaining U.S. military interest along with bad economic depression conditions which reduced existing sponsorship support forced him to return to academia. Ironically, only twelve days after the first successful GIRD launch, Goddard received a letter from the Acting Navy Secretary stating:

> *Because of the great expense that would be entailed in development of the rocket principle for ordinance and aircraft propulsion, which under present stringency of*

funds appears hardly warranted, the Department regrets it is not in a position to further such development.[10]

Goddard received a similarly discouraging rejection letter seven years later in 1940 from the U.S. Army Air Corps. A letter from Brigadier General H. Brett stated:

> *While the Air Corps is deeply interested in the research work being carried out by your organization under the auspices of the Guggenheim Foundation, it does not, at this time feel justified in obligating further funds for basic jet propulsion research and experimentation.*

Goddard tended to eschew publicity, sharing many of his most imaginative ideas only with trusted friends and groups. He did, however, publish a March 1920 letter to the Smithsonian which discussed possibilities of photographing the Moon and planets from rocket-powered fly-by probes, sending messages to distant civilizations on inscribed metal plates, the use of solar energy in space, and the idea of high-velocity ion propulsion.

He also proposed the concept of creating an ablative heat shield for spacecraft entering the Earth's atmosphere covered with layers of a very infusible hard substance with layers of a poor heat conductor between designed to erode heat away in the same way as does the surface of a meteor.[11] This is now the standard solution used to prevent spacecraft from burning up due to surface-atmosphere reentry friction.

Those early ideas, which were generally regarded as very radical at the time, drew strongly sensationalized media publicity and criticism. Following a January 12, 1920 New York Times front page story titled *Believes Rocket Can Reach Moon,* a Smithsonian press release reported about a *multiple-charge, high-efficiency rocket,* a next-day follow-up editorial scoffed at Goddard's proposals. Titled *A Severe Strain on Credulity,* it argued, among other disagreements, that:

> *After the rocket quits our air and really starts on its longer journey, its flight would be neither accelerated nor maintained by the explosion of the charges it then might have left. To claim that it would be is to deny a fundamental law of dynamics, and only Dr. Einstein and his chosen dozen, so few and fit, are licensed to do that.*

Then, to add more insult to injury, the New York Times challenged Goddard's understanding of Newton's fundamental laws. Asserting that thrust can't occur in a vacuum it concluded:

> *That Professor Goddard, with his "chair" in Clark College and the countenancing of the Smithsonian Institution, does not know the relation of action and reaction, and of the need to have something better than a vacuum against which to react—to say that it would be absurd. Of course, he only seems to lack the knowledge ladled out daily in high schools.*[12]

While Goddard finally did get some funding from the U.S. Navy in 1941 for development of rocket-assisted aircraft takeoffs, his work up to that time principally involved guidance and control mechanisms for sounding rockets with few improvements for engine design.

Accordingly, while his rockets never achieved great altitudes, that wasn't really his goal. Rather, his work concentrated upon perfecting liquid-fueled engines along with reliable and accurate guidance and control subsystems which would eventually achieve high altitudes without tumbling in the thin atmosphere and to provide stability for sensitive experiments and other payloads those rockets would carry. By contrast, a German A-4 advanced version of their V-2 reached the outer limits of the atmosphere in 1942.

Although Goddard lacked budgets allocated by the German government, he was on the verge of developing larger rockets capable of reaching extreme altitudes when World War II intervened to change everything.

And yes, such devices later proved to work very well in the vacuum of space after all.

Chapter Four: Von Braun: Germany's Peaceful Space Warrior

BY THE LATE 1930s, when some ballistic missile development was occurring at the U.S. Jet Propulsion Laboratory, the Germans were already developing plans for a major Peenemunde Army Research Facility for fearsome V-2 rocket production in a small town located on Usedom Island on their northern coast. Those activities would ultimately be directed by a charismatic and effective engineer...Wernher von Braun.

Wernher had several characteristics in common with his Soviet counterpart, Chief Designer Korolev. As also with Goddard, both began their careers experimenting as rocket amateurs. Both maintained spaceflight to the Moon and planets as key goals. Both received early funding for military missile development. And, although terms of punishment were vastly different, both were imprisoned for alleged subversion of those military projects.

Unlike Korolev, who suffered hard labor at the notoriously brutal Siberian Kolyma camp, following Germany's WWII defeat von Braun was incarcerated by his U.S. captors under incomparably more comfortable conditions in Fort Bliss, Texas for a mere two weeks. Following his release along with 126 of his former Peenemunde colleagues, von Braun rose to a level of deserved international fame that Korolev could never dream of. Accomplishments of the American-German team he led included development of the Jupiter intermediate-range ballistic missile; the Redstone rocket that launched America's first satellite and first U.S. astronaut Alan Shepard; and the Saturn V rocket that enabled 12 fellow Earthlings to walk on the Moon.

Wernher von Braun was born of a noble family on March 23, 1912 in Wirstiz, Poland, which was at that time part of Prussia and the German empire. As a child he was introduced to astronomy through his mother's gift to him of a telescope. He later became fascinated by speed records established by Max Valier and Fritz von Opel in rocket-propelled cars.

Von Braun's rocket interest was intensified upon reading Transylvanian rocket pioneer Herman Oberth's 1923 book *By Rocket into Planetary Space* (English translation). In 1930 he enrolled at the Technische Hochschule Berlin, earning a bachelor's degree in mechanical engineering in 1932. While there, he joined the Spaceflight Society and assisted Willy Ley in liquid-fueled rocket motor tests in conjunction with Oberth.

Wernher credited Oberth as an important career mentor, stating:

> *Herman Oberth was the first, who when thinking about the possibility of spaceships grabbed a slide-rule and presented mathematically analyzed concepts and designs...I, myself, owe to him not only the guiding-star of my life, but also my first contact with the theoretical and practical aspects of rocketry and space travel. A place of honor should be reserved in the history of science and technology for his ground-breaking contributions in the field of astronautics.*[13]

Von Braun subsequently pursued a doctorate in physics at the University of Berlin, graduating in 1934. That same year, his academic group launched two rockets reaching between one and two-mile altitudes. His graduate studies included rocketry research conducted at a solid-fuel rocket station not far from Berlin under supervision of then-Captain Walter Dornberger, a department head for the German armed forces Ordinance Department.

During this time period the National German Workers Party (NSDAP, or Nazi party) came into power and moved rocketry into the national agenda. His 1934 thesis titled *Construction, Theoretical, and Experimental Solution to the Problem of Liquid Propellant Rocket* was kept classified by the German government and not published until 1969.

In the early 1940s, von Braun moved to the new Peenemunde facility as its technical director under the command of Captain Walter Dornberger where his group, in combination with the Luftwaffe, developed liquid-fuel rocket engines for aircraft and jet-assisted takeoffs. Even more significantly, Peenemunde became the development center for a new A-4 ballistic missile, which became better known as the V-2. The "V" stood for vengeance weapon.

Adolph Hitler recognized the importance of using the A-4/V-2 for military purposes soon after Germany invaded Poland to start World War II in 1939. As reported in a biography published by the Bibliography.com website, von Braun was briefly imprisoned on espionage charges for resisting an attempt by Gestapo Chief Heinrich Himmler to take control of the V-2 project, and he was released under Hitler's personal order.[14]

Von Braun admittedly drew upon Goddard's work taken from various journals for developing Germany's A-4/V-2 rocket. Commenting on Goddard designs, he observed:

> *His rockets…may have been rather crude by present-day standards, but they blazed the trail and incorporated many features used in our most modern rockets and space vehicles.*[15]

A severe labor shortage in 1943 prompted Chief Engineer Arthur Rudolph at the Peenemunde V-2 rocket factory located at Mittelwerk to adopt SS General Hans Kammer's plan to use slave labor. As with other slave labor operations, brutal treatment of worker prisoners produced many tragic casualties. Although von Braun admitted visiting the plant on many occasions and called conditions there repulsive, he claimed never to have witnessed any beatings or deaths directly. However, he admitted that by 1944 it had become clear to him that these incidents had, in fact, occurred.[16]

On December 22, 1942, Adolph Hitler signed an order approving mass production of the V-2 to target London. British and Soviet intelligence agencies soon became aware of the program. Over the nights of August 17 and 18, 1943, the RAF Bomber Command's Operation Hydra dispatched 596 aircraft which dropped 1,800 tons of explosives on the Peenemunde facility. Although it was later salvaged and most of von Braun's team escaped unharmed, the raids killed his engine designer and chief engineer and succeeded in interrupting the program.[17]

Historian Michael Neufeld quotes von Braun in his book *Wernher von Braun: Dreamer of Space, Engineer of War* expressing unhappiness upon hearing news of the London raids. Representing his interests in rocket applications for space travel rather than war, he reportedly said *the rocket worked perfectly, except for landing on the wrong planet.*[18]

Following the end of WWII in 1945, von Braun and his rocketry team (including his brother Magnus) voluntarily surrendered to American forces as part of Operation Paperclip, eventually becoming technical director of the U.S. Army Ordinance Guided Missile Project in Huntsville, Alabama as well as director of the NASA Marshall Space Flight Center from 1960-1970. He also later became vice president of the aviation company Fairchild Industries, Inc. and National Space Institute founder.

Von Braun, a strong life-long advocate for human space travel, missions to Mars in particular, received the U.S. Medal of Science in 1975.

We haven't heard the last of him in this story.

Part Two: The Space Race

ALTHOUGH THE GERMAN V-2 program was not generally regarded to have been a decisive factor in the WWII outcome, with all of those rockets launched roughly equivalent to a single typical RAF raid of perhaps 700 bombers, each dropping two tons of bombs, the remarkable technology caught the keen attention of the USSR and America's allies.

A 1992 Russian Izvestia newspaper interview with former Korolev design bureau member Boris Chertok recounts a secret July 13, 1944 message (before the war ended) from British Prime Minister Winston Churchill to Joseph Stalin requesting permission for English specialists to go into Poland in order to investigate the V-2 test range located in the region of Soviet attack forces. Churchill recognized that Germany's new rocket weapon might pose a serious threat to London.[19]

The Soviets already knew at that time about the existence of a much smaller and more primitive German jet-powered V-1 buzz bomb which was judged not powerful enough to represent great concern. On the other hand, their own investigators in Poland had recovered fragments of an A-4/V-2 which the Germans had unsuccessfully attempted to completely destroy. Close inspection of those remnants caused major alarm.

As reported by Chertok, Soviet rocket specialist Viktor Bolkhovitinov—who examined the remnants—exclaimed upon being asked for an opinion 'What is this?', responded 'This is what cannot be.'

Chertok went on to observe:

> *Understand, one of our most talented aircraft designers simply did not believe that in wartime conditions it would be possible to develop such a huge and powerful rocket engine. We had at the time liquid engines for our experimental rocket planes with thrusts of hundreds of kilograms. One- and one-half tons was the limit of our dreams. Yet here we calculated, based on the nozzle dimensions, that the engine thrust was at least 20 tons.*

Chertok reported that Russians were also shocked to discover that the German rocket was fueled by alcohol and liquid oxygen technology rather than by the customary nitric or kerosene they used. With a gross weight of 12.52 tons, the V-2's advanced turbopump engine design, powered by an 80 percent hydrogen peroxide steam generator, could deliver a 1-metric-ton payload 200 km.

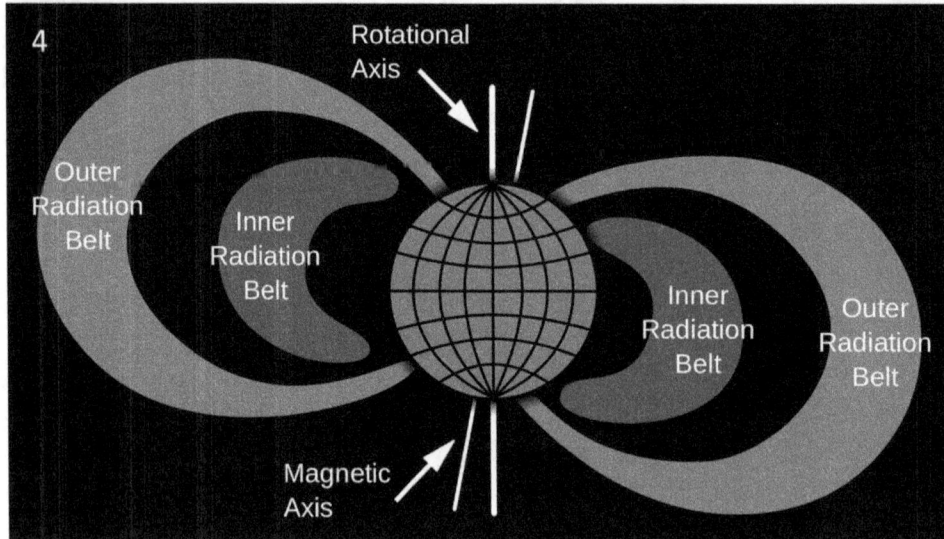

1. V-2 rocket engines at the Mittlewerk manufacturing plant 1945. Source: v2rocket.com
2. Baikonur cosmodrome Source: Former USSR
3. Cape Canaveral Source: space.com
4. Van Alan Radiation Belt Source: Public domain

1. Atlas V
   Source: NASA
2. Navaho Cruise
   Missile
   Source: NASA
3. RD 180 engines
   Source: NASA
4. Saturn V
   Source: NASA
5. Luna -1
   Source: Former
   USSR

Wernher von Braun with F1 engines
Source: NASA

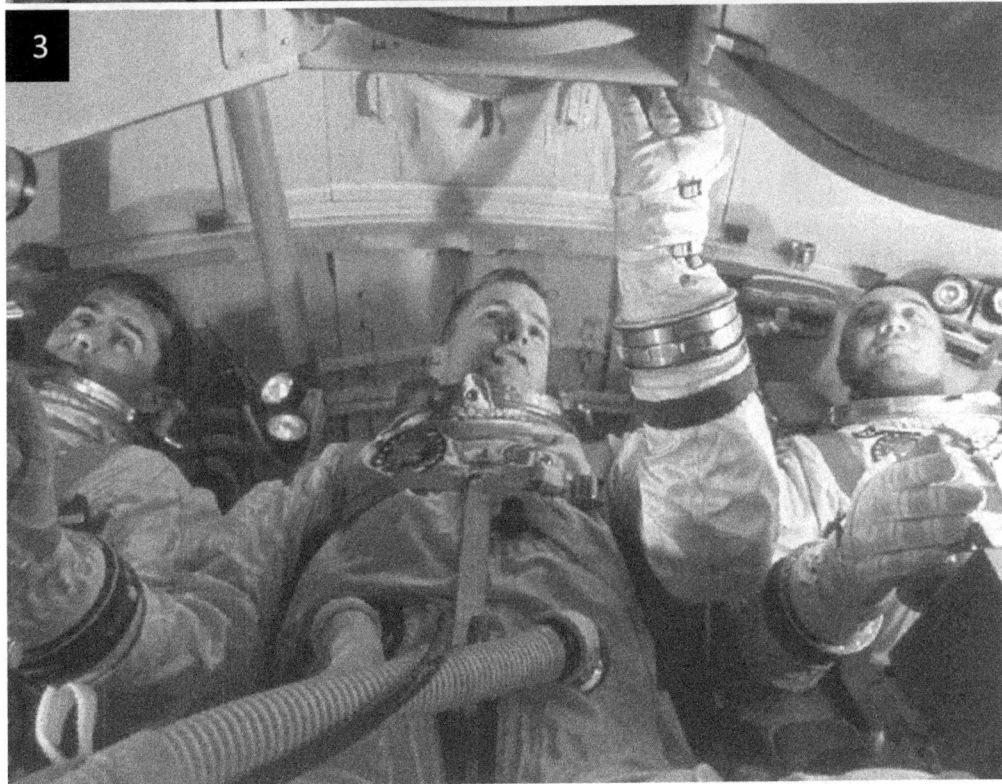

1. John Glenn
   Source: NASA
2. Yuri Gagarin
   Source: ESA
3. Astronauts Grissom, White, and
   Chaffe who were lost in the Apollo
   1 fire in 1966.Source: NASA

1. Chimpanzee Ham
   Source: savethechimps.org

2. Vostok Spacecraft
   Source: Former USSR

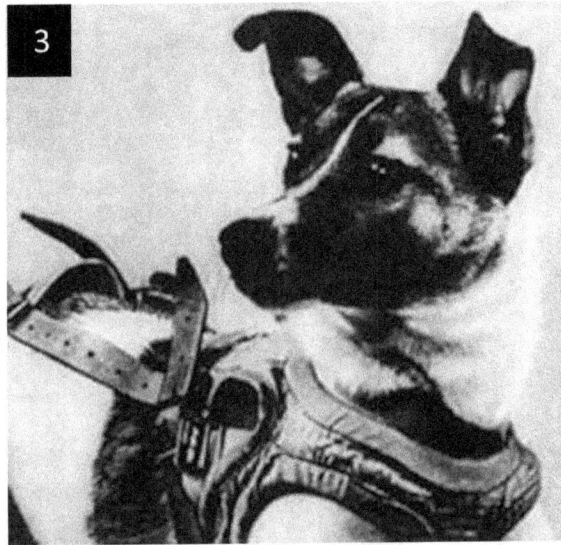

1. Belka and Strelka stamp
   Source: Former USSR
2. Zvyodochka
   Source:Former USSR
3. Laika
   Source: Former USSR
4. Chernushka
   Source: Former USSR

1. Nikita Krushchev
   Source: Dutch National Archives
2. Kennedy at Rice University
   Source: NASA
3. Lyndon Johnson
   Source: Public domain
4. Dwight Eisenhower
   Source: Public domain
5. James Webb
   Source: NASA
6. Alexei Leonov
   Source: Space.com

Chapter Five: Post-War Competition for German Technology

THOSE GERMAN V-2 rocket technology accomplishments weren't immediately credited as significant by Soviet leadership who reportedly expressed a view that whereas America got von Braun and his top leaders, 'We sent our Germans back after a few years.'[20] History, however, tells a different story. Whereas—thanks to von Braun's insistence on keeping his Peenemunde team together—America got the best of that experience, the USSR undertook concerted efforts to round up as many remaining V-2 technicians, blueprints and as much scavenged hardware as could be found in East Germany's Soviet zone.

Soviet V-2 reconstruction and analysis work was begun in a three-story East German electric power station located in Bleicherode near Nordhausen by an unofficially formed organization called the Institute Rabe. Headed by Boris Chertok, that activity eventually involved about 1,000 people, approximately half Russian and half German workers, plus between 50-60 Peenemunde technical veterans.[21]

According to a CIA report later made public in 1960, American intelligence specialists were keeping eyes on Rabe activities to "recreate the design data, test equipment, drawings, manufacturing jigs and tools, ground support equipment, and operating instructions" for the A-4/V-2.[22] The report also observed that the Soviet plans were to set up an assembly line to build the rockets from surviving components, test the engines, build two complete railroad trains for transporting the support equipment and staff for launches, instruct one Soviet and one German launching and testing crew, and develop proposals for further improvement of the A-4.[23]

The Soviets were watching Americans also. Chertok told Izvestia that the Soviets were so interested in knowing what Operation Paperclip was about that they placed agents in the U.S. zone with plans to capture von Braun. That never happened because he was too closely guarded.[24]

Chertok reported that the Russians were awed by the expansiveness of the Mittelwerk V-2 manufacturing plant built under the Kohnstein Mountain near Nordhausen, where the rockets had been moved by Germany after Allied planes bombed the Peenemunde facility. Still, rather than use the gloomy underground site, Russia decided to relocate the test operations to a nearby V-2 repair facility called Klein Bodungen where missile assembly was completed. The original Peenemunde engine development facilities and test stands in the Frankenwald Mountains at Lehesten were later shipped to the USSR following successful firing of a V-2 engine on September 6, 1945.

In addition to these previously mentioned sites, Soviet V-2 technology investigations also took place in many other locations within their zone. Included were a missile controls plant in Berlin, a design office in Sommerda near Erfurt; a Montania engine assembly plant near Nordhausen; and an electrical equipment factory in Sondershausen.[25]

Meanwhile, by October 1945, the American von Braun team was launching a series of V-2 demonstrations as well. Organized by the British near Cuxhaven on the North Sea, the firings were expedited in recognition of the fact that Nordhausen where the rockets were manufactured would soon be in the Russian zone. Scavenging enough weapon material and launch apparatus to fill 200 trucks and 400 freight cars, they managed to assemble only eight complete missiles and launch three. The operations involved twenty-five veteran Peenemunde specialists along with about 1,000 other German nationals including 274 war prisoners plus approximately 2,500 British military personnel.[26]

In addition to shipping enormous quantities of German V-2 equipment items to Russia, the Soviets transported as many as 5,000 skilled German technicians and their families there as well…and not always on a voluntary basis. Those included were not all rocket specialists, but also engineers, scientists and technicians experienced with weapons systems, aircraft and submarine development.

Yet, despite lack of choice in the matter, many of the forced migrants received much better living accommodations and economic circumstances than their Russian counterparts who still suffered wartime privations. This was particularly true for top level experts living in the vicinity of Moscow who gained far nicer housing and two or three times the salaries of their native neighbors. Others of lower technical status were typically sent to modest homes located in various villages near Podlipski, about a forty-mile drive northeast of Moscow.

Most of the Russian and German rocket specialists, including Chertok, reported to a newly created Scientific Research Institute-88 (NII-88) established by the Soviet Party and Council of Ministers showing that missile development was now a national priority. The implementing order was signed personally by Stalin on May 13, 1946 which Russian space historian Maxim Tarasenko characterized as the birthday of the rocket industry of the Soviet Union.[27]

Headed by Lev Gonor, former director of a large artillery plant, NII-88 was divided into three branches, one for conducting experiments, a second dedicated to specialized systems and materials, and a third a Special Design Bureau (SKB) devoted to long range missile development. Befitting the special importance of the SKB, its director was to be a person of special importance as well. The secret name of that individual, a freshly released Soviet prisoner still technically under conviction for counterrevolutionary activity, was Chief Designer Sergei Pavlovich Korolev.

It should be pointed out that Soviet specialists dominated NI-88 activities. Of approximately 150 Germans who worked at NII-88—which included 13 professors, 32 doctor-engineers, 95 diploma-engineers and 21 engineer-practitioners—only 17 were actual Peenemunde veterans. The Germans were organized into work sectors, each supervised by a leader from each country. And although the facility was guarded by woman gunners emplaced at stations along the barbed wire which surrounded NII-8, Boris Chertok reported that this most likely wasn't particularly troublesome for the Germans who had lived under Gestapo conditions.[28]

As reported in James Hartford's book *Korolev: How One Man Masterminded the Soviet Drive to Beat America to the Moon*, one of the young Russian engineers who later became one of Korolev's lead designers recollected:

> *The Germans were working with us cheek by jowl…They showed us how the V-2 was put together. There were ten or fifteen of them where I worked, not hundreds. They lived in a rest home, then later on an island* [Gorodomliya]. *They were delivered to the office by bus every morning. They got twice the money that we got and were allowed to visit Moscow. I learned practical skills from them. We assembled several V-2s, and SAMs* [Surface-to-Air Missiles], *from the hardware brought from Germany.*[29]

Refinements Take Spacecraft to New Heights

NII-88's new Chief Designer Korolev wasn't enamored with the notion that some of his own ideas had not only been realized by the Germans during his imprisonment but had been advanced beyond his most ambitious plans. On top of this was the humiliation of now being tasked to further test embodiments of German concepts rather than to advance his own.

Korolev and his colleagues believed that the V-2 was already obsolete. Leading designers in his group were convinced that they could create a more reliable rocket with a longer range. Yet upon hearing Korolev express those thoughts in a private meeting, Stalin's response was that work must first be completed on the German rocket technology investigations. Those studies led to a launch on October 18, 1947 of the first Soviet-refurbished A-4 developed under German tutelage. However, delayed firing due to repeated failures of igniters embarrassed high level officials in attendance.

The second A-4 demonstration was even less successful. Early in powered flight a guidance gyroscope error attributed to vibration caused the missile to deviate 112 miles (180 km) off course. Following small modifications, a third launch succeeded.[30]

According to Boris Chertok, Korolev asked the triumphant German participants to remain at the site after the successful launch for a talk, telling them:

> *The Minister ordered me to investigate whether the failure of the first two test launchings could have been caused by sabotage of the German specialists. It makes me very suspicious, he said, because the error could be corrected in such a very short time.*[31]

Eventually, eleven of the German-built, Russian refurbished rockets were launched, with five reaching their targets. Meanwhile, the Russians began to produce their own designs for a new R-1 program, where several technologies were developed from scratch at thirty-five research stations and design bureaus along with sixteen principal plants. By 1938, NI-88 had built twelve R-1s, of which seven of nine launched hit their targets.

Within ten years NII-88 had created a series of rockets progressing from the R-1 first commissioned in 1948 to the R-7, the USSR's—and the world's—first Intercontinental Ballistic Missile (ICBM) which is still in service as a space launch vehicle. Its 375 miles (600 km) range was twice that of the V-2/R-1, plus provided a warhead separation capacity to propel that payload rather than the entire vehicle to a target.

Meanwhile, the original German-Russian group submitted a design to the NII-88 science and technical council for a G-1 vehicle no larger than the V-2 to match the 375 miles (600 km) R-7 range with ten times the accuracy. While the proposal failed to move forward due to lack of NII-88 capacity to simultaneously pursue two such major projects, German technicians did accomplish other productive works including development of an improved rocket positioning gyroscope and a quartz clock with accuracy approaching the best mechanical clocks known at the time.

Obviously drawing upon very competent spies, a very detailed 1960 CIA report noted that although most of the Russian team's work was aimed at improving the A-4 performance, some presented new and radical concepts.[32]

According to the CIA report, in spring of 1949 Soviet Minister Ustinov asked the Germans to design a missile capable of delivering a 3-ton warhead 1,550 miles (2,500 km).[33] They responded with two concepts. A proposed R-14 was to be a cone-shaped vehicle about 82 feet (25 m) long and 12 feet (3.7 m) in diameter with a 100-ton thrust. A two-stage R-15 concept would mount a 40 feet (12 m) long cruise missile powered by a winged ramjet second stage to an improved R-1 which would launch it to an altitude of 85-125 miles (140-200 km).

In summer of 1949, senior NII-88 administrator, General Spiridonov informed the Germans:

> *Your work on the R-14, with the exception of some details, will now be terminated. From now on you will work on a new project.*

He went on to explain that they were to develop an anti-rocket rocket capable of encountering and destroying approaching rockets at sufficiently high altitudes to avoid any damage on the ground.

He then reportedly asked:

> *Are there Soviet instruments that allow us to locate approaching rockets precisely and at an early time? And how much time do we have between having located the*

oncoming rocket and its destruction?[34]

The 1960 CIA report observes a policy which emphasized involvement of recent university graduates and even undergraduates in practical on-the-job engineering tasks. Unlike the U.S. where other than a few summer NASA internships, undergraduates typically have little or no exposure to industry prior to being hired, it was Korolev's priority to integrate university education of young engineers in his design bureau.

This wise policy continues.

Beginning in January 1952, the Germans at NII-88 were rapidly being returned to their Fatherland. By this time, the Russians who had milked them of their expertise and technology were well on their way to higher program trajectories without need of further German assistance.

Chapter Six: The Red Star Rises

TO FULLY APPRECIATE Russia's rapid progress in developing ICBM rocket launch, staging, control…and later, atomic warhead armaments and intercept capabilities, one must also consider that these achievements occurred under desperate postwar economic and social conditions.

In 1946, the NII-88 had set up shop in an empty, run-down former military factory where equipment boxes served as desks and design tables, the heating didn't work when it was cold, and leaking roofs caused puddles on the floor when it rained. Most workers depended on food from kitchen gardens commandeered from any land they could find around the nearby countryside, medical care was scant to treat rampant diseases, and sparse housing accommodations were supplemented by dilapidated barracks and tent camps.

Nevertheless, revealing impressive technological capabilities despite these enormous handicaps, beginning from scratch, Russia developed and exploded an atomic bomb in August 1949 just four years after the first U.S. demonstration at Alamogordo. That was followed by the detonation of a hydrogen bomb on August 12, 1953, less than a year after the U.S. did.

On August 21, 1957—four months after Sergei Pavlovich Korolev was notified of an official Khrushchev government ruling that he had been imprisoned unjustly—his team launched a dummy warhead aboard its R-7 ICBM over a distance of 4,000 miles from Russia's important site at Baikonur located in the Republic of Kazakhstan to the Kamchatka peninsula. That was fifteen months before an American Atlas rocket carried a simulated H-bomb from Cape Canaveral more than 6,000 miles over the South Atlantic Ocean on November 28th.

Many distinguished American and Russian engineers had previously argued that cruise missiles afforded more feasible prospects for carrying warheads over long distances than ballistic missiles. News in the 1950s that North American Aviation in California was developing a Navaho cruise missile with a 5,500-nautical-mile (10,100 km) range capable of carrying a 15,000-pound (6800 kg) payload reportedly prompted Korolev to begin working on a similar project. He apparently abandoned that activity upon hearing that America was pursuing the ballistic option. That decision was supported by rapid test successes of their R-7.[35]

Nevertheless, cruise missile development in both countries led to important contributions to ballistic ICBM technology progress as well. Included were advancements to rocket engine design, aerodynamics, materials, and navigation systems.

Ironically, America's present-day Atlas V rockets—which began development as ICBMs intended to deter a USSR threat—now use powerful Russian-designed RD-180 engines that produce nearly one million pounds (nearly 4,500,000 N) of thrust. Also ironically, Wernher von Braun didn't believe the Atlas rocket's thin-walled stainless steel design would survive launch stresses. He referred to it as "no more than a blimp." Whereas the empty weight of the Russian R-7's rugged structure was 23 metric tons, Atlas's was only 7.7 tons—including a warhead—this lower mass offered a big performance advantage.

Spy Games Add to Cold War Chill

Nikita Khrushchev wasted no time pointing a nuclear finger at the U.S. when on November 22, 1957, three months after the R-7 simulated warhead launch, in an interview with publisher William Randolph Hurst, Jr., he threatened:

> *The Soviet Union possesses intercontinental ballistic missiles. It has missiles of different systems for different purposes. All our missiles can be fitted with atomic and hydrogen warheads. Thus, we have proved our superiority in this area.*[36]

Even more pointedly, Khrushchev disparaged the defensive potency of U.S. naval power in a September 7, 1958 letter to President Eisenhower, stating:

> *In the age of nuclear and rocket weapons of unprecedented power and rapid action, these once formidable warships are fit, in fact, for nothing but courtesy visits and gun salutes, and can serve as targets for the right type of rockets.*[37]

The issue of a purported U.S.-Soviet missile gap entered the 1960 presidential campaign when candidate Kennedy claimed that the Russians had built up a substantial lead. President Eisenhower, who had access to spy satellite and secret U-2 overflights, knew differently, but refrained from saying so to protect those sources. Early 1960 CIA Russian ICBM estimates put the total at 35 missiles, with the number growing from between 140 and 200 by mid-1961…not the thousands that some were wildly speculating.[38]

There should be no doubt that both countries were aware they were being spied upon from overhead by the other. Korolev's design bureau had developed and launched four spy satellites beginning with Kosmos-4 (later known as Zenit) on April 26, 1962. By that time the U.S. had already been flying its own spy satellites for nearly two years under a top-secret Corona program that was declassified in 1995.[39]

Zenit was much larger than the American Corona satellite, weighing 4,600 kilograms compared with 850 kilograms. Its cameras could cover the entire U.S. in about twenty-five orbits. At an altitude of about 125-250 miles (200-400 km) it could reportedly determine the number of cars in a parking lot…considered ideal for ICBM site mapping and monitoring. Corona could presumably do the same.

Revelations of Soviet ballistic missile advancements and geopolitical implications made international headlines in 1962 when intended placements in Cuba of R-16 ICBMs triggered a fearsome Kennedy-Khrushchev confrontation. Their range of about 1,370 miles (2,200 km), combined with their basing in Cuba and western Russia, posed a major threat not only to the U.S., but also to bomber bases in Europe and Asia.

The R-16 became the first truly successful Russian ICBM largely owing to advantages of an improved guidance system and a lighter nuclear warhead. Its deployment began in 1961, and by 1965 190 had been installed in different places…some in silos and some in coffin or surface launchers.

That development came with a catastrophically tragic history. An October 24, 1960 R-16 explosion during preparation for firing at Baikonur killed an estimated 165 members of launch and design teams along with several high officials. A particularly notable casualty was Marshal Mitrofan Nedelin, who headed Soviet strategic rocket forces.

The event was covered up from the public, with Nedelin's death attributed to an airplane crash.

A Chirp Heard Around the World

Chief Designer Korolev's responsibilities for ballistic weapons shifted to other priorities in mid-1950s, priorities that would come to change the course of history in big ways.

An October 4, 1957 front page New York Times headline in half-inch capital letters carried a story that was being reported all over the world:

> SOVIET FIRES EARTH SATELLITE INTO SPACE; IT IS CIRCLING THE
> GLOBE AT 18,000 MPH; SPHERE TRACKED IN 4 CROSSINGS OVER US.

That orbit repeated more than 1,400 rounds before Sputnik-1 stopped chirping out its ominous presence and burned up in the atmosphere three months later.

A Le Figaro banner in France announced that "Myth has become reality: Earth's gravity conquered," highlighting "disillusions and bitter reflections" of "The Americans [who] have little experience with humiliation in the technical domain." [40]

Or as an October 7th Manchester Guardian editorial titled *Next Stop Mars* described the event:

> *The achievement is immense. It demands a psychological adjustment on our part towards Soviet society, Soviet military capabilities and—perhaps most of all—to the relationship of the world with what is beyond.*

It went on to speculate that:

> *The Russians can now build ballistic missiles capable of hitting any chosen target anywhere in the world.*

The concept of launching satellites to orbit wasn't new. Konstantin Tsiolkovsky's 1903 calculations showed that a device launched at a certain velocity could overcome the pull of Earth's gravity and achieve orbit. Slightly more than a half-century later, Korolev, another Russian, was now credited with developing the first launcher to achieve that necessary 18,000 mph (8,000 m/s) orbital trajectory.

That goal had been studied and pursued in America as well. A document captured from the Peenemunde Germans motivated the basis for an unsuccessful 1945 proposal by Robert P. Haviland for the U.S. Navy's Bureau of Aeronautics to study the subject.

At about that same time, the (then) U.S. Army Air Force asked major air frame companies to submit secret competitive proposals for design of an Earth-orbiting satellite. A $1 million contract awarded to Douglas Aircraft (then a lot of money) was later switched to a newly formed Project RAND (Research and Development) in Santa Monica, California which produced a report on *Preliminary Design of an Experimental World-Circling Spaceship*. The conclusion predicted that:

> *The achievement of a satellite craft by the United States would inflame the imagination of mankind, and would probably produce repercussions in the world comparable to the explosion of the atomic bomb.*

During early 1954, the U.S. began considering plans to place a small satellite in orbit as its 1957-58 International Geophysical Year (IGY) contribution. In response, Wernher von Braun's team at the Army's Redstone Arsenal in Huntsville began meeting with George Hoover of the Office of Naval Research to accomplish this goal using existing Army Ordinance weapons technology. Their proposed solution, *Project Orbiter,* was to be an Army-Navy-Air Force design. [41]

The Eisenhower administration, however, didn't want the military to get involved in IGY, officially stating that it should be a purely scientific undertaking. Another suspected intent for this policy was to be consistent with the well-publicized Open Skies stance at a time when U-2s and spy satellites were being developed to reconnoiter the USSR. It was reasoned that accomplishing a satellite placement under an umbrella of the IGY program would be less likely to disturb Nikita Khrushchev's sensitivities about overflights since Soviets were also expected to launch a 1957-58 IGY satellite.

In 1955, the U.S. government's guided missile policy committee selected a Project Vanguard to launch an IGY satellite under the auspices of the National Science Foundation using a sounding rocket developed by the Naval Research Laboratory. The design plan used the Martin *Viking* as the basis for the first stage but was an entirely new design. The second stage was developed from the AeroJet General *Aerobee*-but was also a new design.

The Soviets were concerned about America launching an IGY satellite before they did. Towards the end of 1953 Chief Designer Korolev drafted a decree for the Central Committee of the Communist Party that included that possibility based upon a proposal by his visionary friend Mikhail Tikhonravov five years earlier. Then, by 1954, his team's R-7 rocket—with a capacity to carry a 5-ton ICBM warhead—could easily orbit a 1.5-ton satellite. The Council of Ministers authorized Korolev's plan on January 30, 1956 with no time to spare for necessary IGY satellite development preparations.[42]

But what sort of satellite? The Soviet Academy of Scientists was presented with various options for IGY. One possibility was a live organism such as a dog. Another was to fly around the Moon and photograph the side hidden from Earth. The big priority, however, was to beat the Americans. Too ambitious of a plan would fail that purpose.

They finally settled with a plain polished metal sphere carrying only a radio transmitter, batteries, and temperature-measuring instruments. Korolev arranged for Tikhonravov to join his team to develop the concept at Special Design Bureau 385 in the Ural Mountains. It worked, and as a Pravda headline proclaimed in 1957, *World's First Artificial Satellite of Earth Created in Soviet Union.*

Making matters worse here in America, one month later, on December 6th, the first Vanguard launch attempt ironically designated TV-3 (for Test Vehicle 3) failed before world television cameras. Although originally scheduled only as a test, Soviet publicity pressure resulted in moving it up several months. After rising but a few feet off the ground, it ignominiously sagged back, buckled, burst into flame, and tossed its tiny, still-transmitting three-pound (1.3 kg) Sputnik satellite a short distance away. Pravda reproduced a front-page London Daily Herald photo showing the explosion with a superimposed headline which in translation read *OH, WHAT A FLOPNIK!*

Far more fortunately for the U.S., following a Sputnik-2 launch, a von Braun group from the Army Ballistic Missile Agency developed a Jupiter C launch vehicle which successfully placed a 28-pound (13 kg) Explorer-1 satellite in orbit on January 31, 1958. However, it is important to point out that the Jupiter C program was in existence for almost two years and could have launched the first satellite in 1956, if von Braun had been allowed. The scientific benefits proved historic when its onboard instruments first discovered the now famous Van Allen radiation belts.

Although not learnt until many years later, technological bragging rights turned on the Russians after a launch failure three months later on April 27, 1958, with its 1.3-ton Sputnik-III payload aboard.

And while a successful follow-up launch was soon achieved, its replacement Sputnik-III satellite missed an opportunity to further map the Van Allen belts due to an equipment malfunction which failed to record data when the satellite position was out of direct radio contact.[43]

Probing Human Lunar Mission Possibilities

Perhaps influenced by Goddard or his old friend Tikhonravov, Chief Designer Korolev was a long-time advocate for missions to the Moon, first with unmanned probes, later with humans. He had already proposed the possibility of developing a two-stage version of the R-7 rocket to accomplish in a 1955 paper, two years prior to Sputnik-1. It stated:

> *The level of development of technology achieved, at the present time, makes it possible to carry out rocket flight to the Moon.*

A report coauthored by Korolev and Tikhonravov titled *Most Promising Works on the Development of Outer Space* set forth remarkably bold visions:

- By 1958-60: develop a three-stage version of the launch vehicle used for Sputnik plus a more advanced vehicle and interplanetary technology; starting in 1959-60 conduct studies for new chemical fuels and nuclear rockets good for carrying great weight for satellites and extraterrestrial stations; in 1959-63 conduct studies of space station assembly, including use of carrier rocket housings as finished sections of the station; and in 1960-65 develop closed-loop life support systems and design pressure suits for space operations.

- By 1958-61: develop a 10- to 20-kilogram solar-powered research station for landing on the Moon; in 1960-64 develop a fourth stage for a launch vehicle that will enable orbiting the Moon; in 1958-60 develop a heat shield and reentry system for the first Moon-orbiting spacecraft; and in 1959-60 develop a robotic spacecraft with radio and TV equipment capable of flights to Mars and Venus.

- By 1958-62: develop space power stations for space stations and interorbital apparatus; and in 1959-65 study prospects for new developments on long-distance communications.

- By 1962-66: study spacecraft rendezvous techniques; in 1963-64 develop a new launch vehicle that can put a 15- to 20-ton space station into orbit and permit the possibility of interplanetary flight; in 1961-65 complete design of a two- or three-man space vehicle for prolonged stay in space…and develop an ion engine-powered spacecraft for manned flight around the Moon; in 1963-66 create a robotic space vehicle for flights to Mars and Venus; and beginning in 1962 undertake the design for a space station in orbit for studying the effects of weightlessness on plants and humans and for studying radiation effects on animal and vegetable organisms.

The report concluded:

> *After the realization of these planned works…can be set on the following tasks: manned flight to Mars and Venus, manned flight to the Moon with landing and return to Earth, construction of a permanent station-colony on the Moon (beginning with research in 1960).*

Korolev is believed to have made his first serious proposal for a manned Moon mission during an April 6, 1956 speech to the Soviet Academy of Scientists…more than a year prior to the Sputnik-1 launch.

He stated:

> *This real task is to fly to the Moon and back from the Moon. This task is most easily solved by starting from Earth. Somewhat more difficult will be returning to Earth that will be on a satellite or rocket that goes to the Moon. But it must not be believed that the proposals I am making are extremely remote.*[44]

Korolev's writings made his primary objective clear:

> [These] *first studies of the Moon and interplanetary space at distances that reach 400-500 thousand kilometers will also create the necessary prerequisites/premises for the penetration of man into interplanetary space, the Moon and the planets.*[45]

According to his daughter Natasha, Korolev's passion for reaching the Moon was so great that he called his lunar rockets Mechtas (or dreams)...although they were officially termed Lunas. Since the first three launches failed to reach Earth orbit from which they would be transferred to a lunar trajectory, therefore, according to custom, they never received subsequently recorded Luna-1 designations. A January 2, 1959 launch received an official Luna-1 name because even though missing its target by 3730 miles (6,000 km), it was the first manufactured object to orbit the Sun. In addition, its instruments determined that the Moon had no magnetic field.[46]

Luna-2, launched on September 12, 1959, became the first spacecraft to make contact with the Moon or any other celestial body. Tracking data from stations at Jodrell Bank (UK) and the U.S. Fort Monmouth Observatory in New Jersey confirmed a Soviet claim of that achievement.

Luna-3 launched on October 4, only three weeks after Luna-2, photographed the far side of the Moon never before seen by humans. This feat accomplished one of the ambitious 1957-58 IGY proposals considered by the Soviet Academy of Sciences.

Meanwhile, just as the Soviets had experienced some Moon-shot failures, Americans were realizing even more of our share of problems. These began on August 17, 1958, with a first stage Air Force Thor-Able malfunction of the subsequent Pioneer spacecraft 77 seconds after launch from Cape Canaveral. Although a Pioneer-1 third stage missed the Moon, it set a distance record by traveling some 70,745 miles (113,854 km) into space. Pioneer-3 then provided important data about the outer Van Allen radiation belt.

Notwithstanding these significant achievements, there were seven straight Pioneer Moon misses through 1960.

The Non-Space Race

Needless to say, these discouraging setbacks didn't reflect well on U.S. attempts to initiate a publicly credible space program. A draft portion of Dwight Eisenhower's January 1960 State of the Union speech prepared by NASA's first administrator T. Keith Glennan rationalized the embarrassments, stating:

> ...*we are not going to attempt to compete with the Russians on a shot-for-shot basis in attempts to achieve space spectaculars.*[47]

Reflecting government priorities and sparing no expense, the level of activity at Korolev's design bureau stepped up at a frenzied pace, a particularly remarkable circumstance given the desperate state of the Russian economy. Agendas included new spacecraft developments for the Moon, Mars, and Venus and preparations for a human orbital mission. All of these operations would depend upon various versions of the venerable R-7 launch vehicle outfitted with new upper stages powered by an advanced RO-5 engine. Developed in only nine months, the RO-5 had a 5.6-ton thrust.[48]

Korolev ordered the creation of a new Center for Deep Space Radio Communications located in the Crimea to supplement spacecraft tracking and trajectory corrections during periods when radio contact was lost from the single USSR station due to Earth rotation. A U.S. Deep Space Network developed by 1965 afforded round-the-world coverage with stations in Goldstone, California; the Auroral Valley near Canberra, Australia; and Johannesburg, South Africa.

As moving targets, Mars and Venus destinations presented far more limited astronomical launch windows and more complicated trajectory guidance challenges than the Moon. Two October 1960 Soviet Mars probe failures to reach Earth orbit were followed by seven straight failed Venus probes—five by Russia and two by America—between February 1961 and September 1962.[49]

The first failed Soviet Mars launch which occurred at the time of the Khrushchev-Kennedy standoff over Cuban missile emplacements might very well have led to a little-publicized but hugely larger competitive Russian-U.S. disaster.

Just as the spacecraft was being prepared for launch at the Baikonur pad, Korolev's team was ordered to immediately remove it and abort the mission so that a military ICBM could use the site in response to a U.S. thermonuclear strike. After the issue soon appeared to be settled via diplomatic channels, launch preparations were allowed to proceed.

That didn't end the problem. On October 24, 1962—still in the middle of the crisis, the Mars launcher exploded into so many pieces during ascent that observers at the U.S. Ballistic Early Warning System feared that a Soviet nuclear attack might have commenced. Crisis was averted when computers which assess trajectory and impact points reported a false alarm within seconds.

The international nuclear war scare subsided on October 27, 1962 when Khrushchev announced he would dismantle the missiles in Cuba and return them to the USSR. One week later, a Soviet spacecraft designated Mars-1 made the first (unintentional) Red Planet flyby after losing communications. Nevertheless, it accomplished the impressive feat of traveling 66 million miles (106 million km) and sending back 61 batches of data until March 21, 1963.

The disclaimed U.S.-Russian competition intensified over the next several years, with each nation anticipating and closely monitoring activities of the other. Both continued to experience significant, if incremental, successes and failures.

A Russian Zond-1 spacecraft launched on April 2, 1964 reached the vicinity of Venus on July 20, although a radio failure resulted in no returned data. Venera-3 launched on November 16, 1965 accomplished the first Venus impact on March 1, 1966. A Zond-2 launched on November 30, 1964 demonstrated use of the first electric thrusters for attitude control.

On the American side, although its instruments failed, on April 23, 1962 Ranger-4 became the first U.S. spacecraft to impact the Moon. Launched on October 18 of that year, Ranger-5 missed the Moon by 435 miles (700 km). Ranger-6 hit the Moon on January 30, 1964, but the TV camera didn't work.

Rangers 7, 8 and 9 returned marvelous pictures covering over 150,000 square miles (400,000 square km) of the lunar surface between 1964 and 1965. And while Mariner-1 failed to reach Venus, Mariner 2 flew within 22,000 miles (35,000 km) of the planet. Mars-3 missed Mars, but in July 1965, Mars-4 sent back spectacular TV pictures of its cratered surface from a distance of 6117 miles (9,844 km).

The 1970s witnessed more historic interplanetary achievements by both Russia and the U.S. In 1971, five years after Korolev's death, two Soviet capsules released by Mars-2 and Mars-3 crashed into the Martian surface on November 27 and December 2, respectively. The first of those events occurred just two weeks after the American Mariner-9 developed by NASA's Cal Tech Jet Propulsion Laboratory orbited around the planet throughout a dust storm, then sent back detailed pictures of the surface until January 1972.

Gagarin Puts a Human Face on Space

Although Chief Designer Korolev never lived to witness Russian probes reaching Mars, he did experience a personal triumph which commenced a transformational new era of human space exploration.

As the *New York Times* exclaimed on April 12, 1961, once again in bold front-page headlines:

SOVIET ORBITS MAN AND RECOVERS HIM; SPACE PIONEER REPORTS: 'I FEEL WELL': SENT MESSAGES WHILE CIRCLING EARTH.

The newspaper followed up with an editorial the next day prophesizing that the "flight will be hailed as one of the great advances in the story of man's age-old quest to tame the forces of nature." Pravda declared it a "GREAT EVENT IN THE HISTORY OF MANKIND." The Communist Party seized upon the event as a triumph over capitalism.

This was not generally greeted as good news by the majority of Americans, and particularly not by those connected with the U.S. space program. Gagarin's one-hour, 48 minute full-orbit demonstration aboard a Vostok spacecraft on April 12, 1961 eclipsed a 15-minute-long suborbital launch of Navy jet pilot Lt. Col. Alan Shepard on May 5th of that year which reached a 167-mile altitude and travelled 302 miles downrange. Unfortunately for American history, Shepard's flight was delayed multiple times following a von Braun decision that another test flight was needed after a previous one traumatized its passenger…a chimpanzee named Ham.

Thirty-eight-year-old Shepard was deeply disappointed to have his chance to be the first human to orbit the Earth preempted by the 27-year-old Russian pilot. Shepard, along with another Project Mercury candidate, U.S. Air Force pilot Gus Grissom, had undergone a year of training for the opportunity, whereas Gagarin's preparation had begun only six months before and was notified of being selected just four days prior to launch.

Although the distinction of being the first American to orbit Earth ultimately went to John Glenn, fortune later beamed more brightly upon Alan Bartlett Shepard, Jr. upon becoming the only Mercury astronaut to walk on the Moon on the Apollo 14 mission. Gus Grissom suffered far worse fortunes. He nearly drowned when explosive bolts fired unexpectedly, blowing the hatch off his Liberty Bell capsule during splashdown following the second suborbital Project Mercury-Redstone flight on July 21, 1961. A catastrophic fire ended the lives of Apollo Astronauts Gus Grissom, Roger Chaffee, and Ed White during a January 27, 1967 test.

Comparing NASA and Russian selection criteria reveals a very contrasting orthodoxy. While both drew upon candidates with military aviation backgrounds, NASA selected more experienced test pilots who were typically 10 years older than Russian cosmonauts. Russians picked for cosmonaut training were also smaller than American astronauts, with heights limited to 5'6" (168 cm) and a maximum weight of 143 pounds (65 kg). The first seven U.S. astronauts averaged more than three inches (8 cm) taller and 20 pounds (9 kg) heavier than their Soviet counterparts.

Russian and NASA spacecraft design and operational strategies differed markedly as well. Each set of approaches, partly driven by respective technology maturity status, offered distinct, often safety-related advantages and liabilities.

Beginning with all-important launch and spacecraft control features, the then-existent U.S. Atlas capabilities limited NASA's Project Mercury orbital delivery capacity to one metric ton, whereas the Soviet R-7 could lift a 5-ton capsule. America also adopted a truncated blunt-nosed cone Mercury capsule configuration to reduce shock wave and absorb 90 percent of heat upon reentry from space. By comparison, the Vostok capsule that carried Gagarin was spherical in shape like sputnik. Being spherical, it was dynamically more stable, thereby minimizing a need for complex attitude control devices which were less advanced than America's. It also had its center-of-gravity positioned so that it would become "stable" with the astronaut flying backwards during reentry so as to distribute the G-forces in a "positive' manner across his body.

U.S. emphasis upon seasoned pilots and the Soviet's upon automated control systems was highly evident in their contrasting operational protocols. Unlike the passive role allocated to cosmonauts, American astronauts were provided with controlling means to override automated systems. The significance of this difference nearly led to disaster when Gagarin's spacecraft went into a wild 10-minute-long spin which he was unable to manually correct. This event was held secret until publicly revealed on the 30th anniversary of that flight in 1991.

Still another difference in design philosophies ultimately lead to a disaster, this time on the American side. Whereas the Russians chose for spacecraft cabin pressurization and breathing to approximately match Earth-normal 80 percent nitrogen/20 percent oxygen at one atmosphere, Americans chose instead to provide pure oxygen at one-third atmosphere, reasoning that the normalized nitrogen-oxygen approach would expose astronauts to the bends in an emergency event requiring them to switch to a pressurized space suit during flight.

On the other hand, while a pure oxygen atmosphere wouldn't cause a decompression problem provided that astronauts pre-breathed oxygen prior to flight to remove nitrogen from bloodstreams, an absence of nitrogen in a 100 percent oxygen atmosphere could result in explosive fires. Although these didn't occur during seven Mercury missions (between 1961-1963), or throughout ten Gemini missions (1965-66), the decision led to the January 27, 1967 Apollo test tragedy.

Russia and America had both been conducting animal tests to determine if humans could survive launch and reentry stresses in addition to weightless orbital conditions. Soviet space scientists had been experimenting with canines since at least 1951. Dogs Dezik and Tsygan were sent to 62 miles (100 km) altitude that year using same pod that carried Laika. U.S. had experimented with monkeys since 1950s.

Studies in both countries concluded that weightlessness and high accelerations wouldn't adversely affect humans. These determinations must obviously have influenced the dog launch demonstration proposed for the 1957-58 IGY. After instead deciding to orbit a far simpler Sputnik-1 radio transmitter payload, the idea immediately resurfaced for Sputnik-2.

Not known to Westerners until mentioned in a 1994 publication, Russian space canine experiments had not always led to successes. A July 28, 1960 Vostok prototype flight carrying dogs Chaika and Lisichka failed when the launch vehicle exploded. More fortunately, an 18-orbit flight the following month ended far better for dogs Belka and Strelka who became the first creatures to return alive.

A few weeks later the Communist Party approved a request to launch a human. Not even the Baikonur ballistic missile explosion on October 24, 1960 that killed 165 people including Marshal Nedelin halted plans to go ahead. Nor did the failure of another Vostok prototype which killed two more dogs, Pehelka and Mushka on December 1st when a control error caused the cabin to burn up upon entering the Earth's atmosphere at a steeper angle than planned.

During a problematic December 22, 1960 attempt, dogs Damka and Krasvka were recovered alive after failing to achieve orbit due to a premature third-stage engine shutdown. On two subsequent flights, dog Chernushka successfully flew on March 9, 1961, and dog Zvezdochka was recovered alive on March 25.

As later recounted by Cosmonaut Gyorgi Grechko, Khrushchev summoned Korolev along with others of his team to the Kremlin five days after the October 4, 1957 Sputnik-1 achievement, telling them:

> *We never thought that you would launch a sputnik before the Americans. But you did it. Now please launch something new in space for the next anniversary of our revolution.*

That was only one month away.[50]

With no previous preparations in place, Korolev and his colleagues accomplished that order on an amazingly short schedule. Weighing about 1,000 pounds (450 kg), Sputnik-2 launched the world's first (temporarily) living space passenger, a mongrel dog named Laika, on November 3, 1957 along with a duplicate Sputnik-1 sphere. While the launch was successful, the mission didn't go well for Laika. Although she survived the launch, she soon perished when the capsule overheated after failing to separate from the booster, rendering the thermal control system inoperative.

The two sputnik successes provided Nikita Khrushchev with bragging rights which he featured in a speech at the fortieth anniversary of the Soviet November 6 Revolution:

It appears that the name Vanguard reflects the confidence of the Americans that their satellite would be the first in the world. But...it was the Soviet satellites which proved to be ahead, to be in the vanguard...In orbiting our Earth; the Soviet sputniks proclaim the heights of the development of science and technology and of the entire economy of the Soviet Union, whose people are building a new life under the banner of Marxism-Leninism.

Again, Chief Designer Korolev was kept off to the side away from public fanfare following Gagarin's triumphant return. Worse, NASA achievements were sidelined altogether.

George Low, then chief of manned space flight, recalled a conversation between newly appointed NASA Administrator James Webb and his deputy Robert Seamans who had just testified on the state of their efforts before the House Committee on Science and Astronautics on the day before Gagarin's flight deciding not to show a film following Gagarin's historic world spectacle. The movie featured recovery of the dazed chimpanzee Ham two and one-half months earlier.

Low remembered the conversation, prudently concluding:

...it would not be in our best interest to show how we had flown a monkey on a suborbital flight when the Soviets had orbited Gagarin.[51]

Chapter Seven: America Leaps Forward with Right Stuff

A CHASTENED AMERICA got the message without need of the dazed monkey film. Nor was it the only big humiliation facing President Kennedy's new administration. The Cuban Bay of Pigs debacle had just recently occurred in mid-April as well.

The particularly unfortunate timing of these two events put great pressure on Kennedy to demonstrate resolute leadership. On May 25, 1961, just slightly less than six weeks following Gagarin's catapult into the Space Hall of Fame, he did so, announcing before a special joint session of Congress that an American astronaut would be safely sent to the Moon "...before this decade is out."

In response to a reporter's question regarding when the U.S. might perhaps surpass Russia in this field, President Kennedy providently observed "...the news will be worse before it is better."

On August 6, 1961, Gherman Titov's seventeen-orbit Vostok-2 flight topped Gus Grissom's suborbital Redstone-4 flight of July 21.

Kennedy repeated his commitment to human lunar exploration as a national priority at Rice University's stadium on September 12, 1962, stating:

> We choose to go to the Moon. We choose to go to the Moon in this decade and do
> the other things, not because they are easy, but because they are hard, because that
> goal will serve to organize and measure the best of our energies and skills, because
> that challenge is one that we are willing to accept, one we are unwilling to
> postpone, and one which we intend to win, and the others, too.

The president also made it clear that this was to be a competitive race dedicated to demonstrating U.S. technological supremacy over the Russians:

> Within these last 19 months at least 45 satellites have circled the earth. Some 40 of
> them were made in the United States of America and they were far more
> sophisticated and supplied far more knowledge to the people of the world than
> those of the Soviet Union.

What had previously been a tacit matching of wits and capabilities involving post-war rivals had become an officially recognized race between superpowers. The situation literally began to look up for America when a Mercury-Atlas 6 rocket launch of Friendship-7 carried Colonel John Glenn on three orbits on February 20, 1962.

Three months later, Mercury-Atlas 7 carried Scott Carpenter on three orbits, splashing down 260 miles (420 km) beyond the target area due to reentry errors. On October 3, 1962, Mercury-Atlas 8 Astronaut Walter Schirra did even better, splashing down within 7.24 km of a recovery ship following 6 orbits. That record, in turn, was beaten by Gordon Cooper on May 16, 1963 on Mercury-Atlas 9 which landed 6.4 km from ship following 22 orbits.

Clearly, a space race had begun after all. On August 11-15, 1962, Soviet Cosmonaut Andriyan Nikolayev did 65 orbits aboard Vostok-3, while Pavel Popovich who launched the next day aboard Vostok-4 did 48, the two spacecrafts flying in orbits within 3 miles (5 km) of each other. On June 14-19, Valeri Bykovsky—aboard Vostok-5—passed within 3 miles (5 km) of Russia's (and the world's) first woman, Cosmonaut Valentina Tereshkova, aboard Vostok-6.

A Challenge Taken Seriously

The high international prestige stakes of this competition weren't lost on Khrushchev, Korolev and others in the Soviet Union. As quoted by Korolev associate Oleg Ivanovsky:

> He would tell us that 'the Americans are at our heels, and the Americans are serious people.' He wouldn't use the word 'Amerikantsi' but 'Amerikan-ye' as if these weren't just American residents but the entire American culture we were competing with. He didn't mean this as an insult but as a show of respect for the competition.[52]

Korolev and his design bureau responded with a very cramped three-person Voskhod (meaning rise as in sunrise) capsule which was hastily ordered in 1964. The spacecraft was a three-person version of a Vostok which had been planned from the time Soviets first became aware of America's two-seater Gemini lunar test program in late 1961. However, it was more a political stunt to "leapfrog" the American Gemini by supposedly flying a three-person spaceship well before the American Apollo. The Soviets crammed three cosmonauts in the same space as had been occupied by one, by omitting their space suits and some other additional equipment.

The feverishly rushed Voskhod development schedule imposed very risky safety compromises. To reduce crowding and weight it was decided that cosmonauts would not wear bulky pressure suits inside the cabin. Emergency ejection hatches were also deleted. In addition, there was inadequate time to test performance characteristics of a new pressure suit that would be worn during an excursion outside an orbiting capsule during the second of two total Voskhod missions.

On October 12, 1964—just seven months after the project was approved—Voskhod-1 orbited its three tightly-packed passengers 16 times following a successful unmanned test flight only six days earlier. Then, only five months later on March 18, Cosmonaut Alexei Leonov conducted the world's first space walk from Voskhod-2, one which proved highly stressful.

As with aspects of other hurried preparations to beat the Americans, design and development of the airlock and pressurized suit for Leonov's spacewalk were accomplished under great time pressure with little opportunity for testing.

Rather than install a hatch door that when opened would depressurize the cabin and expose all occupants to vacuum as was done with Gemini, a collapsible, cylindrical, three feet (1 m) diameter chamber constructed of two layers of rubber-like material was attached with pressure locks at both ends. The pressure suit was designed and fabricated within a record short period of nine months.

Although prototypes of both the pressure suit and deployable airlock were launched together for testing about a month and a half before Leonov's flight, results were lost when the unmanned spacecraft exploded due to an erroneous ground station self-destruct signal. There was not sufficient time to repeat the experiment before Voskhod-2 launched.

While not reported until years later, all did not go as planned. Immediately upon stepping out into space,

Leonov realized that the pressure suit ballooned to the point that his hands slipped out of the gloves and his feet came out of his boots. Worse, his misshapen suit would no longer fit through the airlock for reentry. Realizing no alternative, and without informing ground control, Leonov bled about half of the air out of his suit through a valve in its lining. With no time to lose upon experiencing sensations of pins and needles in legs and hands, early signs of potentially fatal bends, he entered the airlock headfirst rather than leading with feet. This required him, blinded with perspiration, to turn around in the tiny space and ensure the umbilical cord which provided air to his suit was entirely inside, then lock the hatch behind him.

After finally reentering the spacecraft, more problems arose. First, the firing of small explosive charges to eject the no-longer-needed airlock into space caused the Voskhod to rotate out of control. Then, even more serious, soaring oxygen levels made the cabin dangerously flammable, subject to explosion in the presence of any spark.

Still another problem occurred when the spacecraft's automatic re-entry system retrorockets failed to fire as scheduled. This required that they be to be fired manually and with great precision, something never previously attempted by Russians. If the rocket burn was too short, Voskhod-2 would hit Earth's atmosphere at too shallow an angle, causing it to bounce back into space. Too long a burn would cause too steep an angle resulting in catastrophic destruction.[53]

Although the manual firing went well, Voskhod-2 overshot its intended landing area by 1200 miles (2,000 km), setting down in the western Ural Mountains in a remote corner of Siberia north of the industrial city of Perm in deep snow and wedged between two large fir trees. The cosmonauts spent a bitterly cold night in the capsule with wolves howling around them, finally skiing to a rescue helicopter which was able to land about 650 feet (200 m) away. They spent a second night in tents and warm clothes provided by rescuers.

Sergei Pavlovich Korolev—who had maintained constant communication with the head of search service—reportedly offered some humorous advice after the rescue was complete. He ordered "And now bring a half kilo of Validol [a tranquilizer] for the State Commission." [54]

The problem-prone Voskhod-2 flight took place only five days before Gus Grissom and John Young became the first U.S. pair to orbit Earth in a Gemini vehicle. The rest is history.

Whereas Kennedy had set a national 1961 goal to put a human on the Moon and safely return him within a decade, America accomplished that and even more. By 1969 America had landed four of our citizens, plus delivered two more into lunar orbit who returned with them. Within three more years, eight others had walked on the Moon on successful round-trip voyages, along with four more Moon-orbiting companions.

Some of these same Apollo astronauts, and many courageous predecessors, literally blazed that pathway. They flew on two suborbital and four Earth-orbital Mercury launches, nine Gemini flights, two Earth-orbital tests, and two lunar-orbital tests that made those lunar surface landings possible.

As Kennedy warned at the time, these accomplishments didn't come easily. They required development of a colossal new launch vehicle, invention and demonstrations of countless scientific and technological enablers, and real-time on-the-go planning and coordination of complex operations spanning the country. The costs were enormous, including the lives of Gus Grissom, Ed White and Roger Chaffee who perished during an Apollo command module ground test that taught a tragic lesson.

Why Russia Lost the Moon Race

First, consider that Russia began with a head start.

NASA didn't even exist at the time of the Sputnik-3 launch on May 15, 1958, an event which first prompted serious congressional and White House interest leading to its creation.

Soon afterwards then-Senate Majority Leader Lyndon Johnson instructed his Preparedness Subcommittee:

> *Start work at once on development of a rocket motor with a million pound thrust...put more effort in the development of manned missiles [satellites]...accelerate and expand research and development programs, provide*

funding on a long-term basis…and improve control and administration within the Department of Defense or through the establishment of an independent agency.

That same year, President Eisenhower, who had previously shown little interest in space, concluded along with his science advisor James Killian that a civilian space agency was needed. Hugh Dryden, head of the National Advisory Committee for Aeronautics (NACA) was tasked to prepare legislation to make that happen. Eisenhower signed Public Law 85-568 on July 29, 1958 which created the National Aeronautics and Space Administration (NASA).

NACA, the core group which formed NASA, had an annual budget of about $100 million which supported about 8,000 employees. While they were only four months behind the Soviets in launching a satellite, they were four years ahead in rocket engine capabilities. In fact, Project Mercury, which launched John Glenn to orbit, became official only five days after NASA was born and was conceptualized by a small group of engineers who had worked for NASA for less than three months.

Keith Glennan, Eisenhower's pick to head NASA, later indicated challenging circumstances facing the fledgling organization. He reflected in his memoirs:

> *The philosophy of the* [Mercury] *project was to use known technologies, extending the art as little as possible, and relying on the unproven Atlas. As one looks back it is clear that we did not know much about what we were doing.*[55]

Three months into his presidency and just two days after Yuri Gagarin was launched into orbit on April 12, 1961 on an R-7 rocket, John Kennedy convened a meeting with James Webb—who had succeeded T. Keith Glennan as NASA director—along with NASA Deputy Director Hugh Dryden to explore ways to catch up with and pass the Russians.

Kennedy emphasized that the United States intended to maintain its position in world leadership, its position of eminence in commerce, in science, in foreign policy, and in whatever else might develop from space exploration.[56]

On April 20, only slightly more than a week after Gagarin's flight and a day after realizing disastrous results of the CIA-abetted invasion at the Bay of Pigs by Cuban exiles, President Kennedy called in Vice President Lyndon Johnson whom he had appointed to chair a newly formed National Space Council for advice.

A memorandum from Kennedy[57] instructed Johnson:

> *…to be in charge of an overall survey of where we stand in space…do we have a chance of beating the Soviets by putting a laboratory in space, or by a trip around the Moon, or by a rocket to land on the Moon and back with a man.*

The memo also asked Johnson to determine:

> *Is there any other space program which promises results in which we could win? [and] …how much additional would it cost?*

Wernher von Braun concluded in an immediate response to Johnson that the U.S. had:

> *…an excellent chance of beating the Soviets to the first landing of a crew* [not just a man] *on the Moon.*

He had good reasons to believe so. In order to get there first, the Russians would need to somehow increase their current rocket launch capacity by a factor of ten times. He was certainly also well aware of several years of work by North American Aviation's Rocketdyne division on a huge F-1 engine.[58]

A three-stage Saturn V rocket using five proven F-1 engines for the first stage offered a formidable U.S. advantage. Standing 363 feet tall (58 feet higher than the Statue of Liberty from ground to torch) with a 260,000-pound (120,000 kg) payload capacity, it would dwarf all previous rocket capabilities within the time allowed. A four-stage Apollo-Saturn rocket would be required to place payloads in lunar orbit.

Chief Designer Korolev was also aware of the Russian advantage, but was unsuccessful in obtaining funding necessary to complete development of an N-1 booster that would be comparable in thrust and size to Saturn. Russia's economy was in shambles, and other priorities, military in particular, took precedent. Following repeated requests to Khrushchev and Brezhnev who followed, neither would release the rubles. The Kremlin's top priorities during the 1960s were thermonuclear warheads, ballistic missiles, and submarines to deliver them. This lack of support was kept so secret that the Kremlin pretended until the end that there was never a Soviet manned Moon program at all. Ultimately historical facts profoundly disproved Nikita Khrushchev and Communist Party attributions of Soviet space achievements as triumphs over capitalism.

Despite a head start, Russia's failure to beat America to the Moon wasn't due to a lack of space leadership or competent engineers. As in America, their programs drew upon an abundance of exceptional people…not only Germans, but home-grown as well. No, what the Russians lacked in order to successfully compete in a Moon race was an economically and technically powerful private industry infrastructure fueled by capitalist incentives. And regardless of whether the Bay of Pigs debacle may have had any influence in setting priorities, America had a charismatic president who rallied popular support for a bold national vision to lift us there.

Referring back again to President Kennedy's May 25, 1961 speech committing the U.S. to send a man to the Moon and return him safely before the end of a decade, he challenged the country to recognize that:

> …no single space project in this period will be more impressive to mankind, or more important for the long-term exploration of space; and none will be so difficult or expensive to accomplish…in a very real sense, it will not be one man going to the Moon—if we make this judgment affirmatively, it will be an entire nation. For all of us must work to put him there.

And there was a remarkable clincher:

> …this is a judgment which the Congress must finally make—let it be clear that I am asking the Congress and the country to accept a firm commitment to a new course of action, a course which will last for many years and carry very heavy costs…if we are to go only half way, or reduce our sights in the face of difficulty, in my judgment it would be better not to go at all.

America chose wisely.

Part Three: International Competition and Cooperation

ALTHOUGH NOT BROADLY known, two years before his assassination on November 22, 1963, Kennedy had proposed a possible joint cosmonaut-astronaut lunar mission to Khrushchev at a June 3, 1961 Vienna luncheon.

After some consideration, Khrushchev rejected the offer.

Kennedy proposed the idea again—just two months before his death—at a September 20, 1963 United Nations General Assembly appearance, stating:

> *Space offers no problems of sovereignty...Why, therefore, should man's first flight to the Moon be a matter of international competition? Why should the United States and the Soviet Union, in preparing for such expeditions, become involved in immense duplications of research, construction, and expenditure? Surely we should explore whether scientists and astronauts of our two countries— indeed of all the world—cannot work together in the conquest of space, sending some day in this decade to the Moon not the representatives of a single nation, but the representatives of all of our countries.*[59]

Understandably, Kennedy's offer to team with Russia in a joint U.S.-Soviet manned lunar program wasn't accepted with any good cheer by his NASA Apollo team who saw things very differently. Some reacted in disbelief. Others viewed it as gamesmanship on the part of Kennedy intended to embarrass America's nemesis.

NASA Administrator James Webb reported at the time that Kennedy advisor McGeorge Bundy:

> *...is quite open to an exploration of possible cooperation with the Soviets and thinks they might wish to use our big rocket, and offer in exchange the advanced technology which we are likely to get in the near future.*[60]

Khrushchev had a different interpretation of the tradeoff. Writing in his memoirs he said:

> *Had we decided to cooperate with the Americans in space research, we would have had to reveal to them the design of the [R-7] booster for the Semyorka.*[61]

In any case, while neither Kennedy nor Khrushchev would live to see their astronauts and cosmonauts joining together on evening Moon strolls, both nations would leave separate tracks. Twelve Apollo explorers—beginning with Neil Armstrong and Buzz Aldrin—would impress human footprints on its surface, while robotic Soviet explorers would leave mechanical trackprints before the end of the Apollo program in 1974.

The Russian Lunokhod (Moonwalker) program—which was originally intended to support manned surface missions—produced remote-controlled rovers that predated all others to be deployed on a celestial body. Launched aboard powerful Proton-K rockets, they were controlled from a network of ten ground-based facilities containing Earth satellite vehicle tracking equipment along with command/controls for Soviet near-space civil and military operations.

The first of the total two successful Lunokhod rover deployments (Luna 17) reached the Moon in the Sea of Rains on November 17, 1970. Measuring 4 ft-5 inches high, it carried four television cameras, special extendable devices to collect lunar soil for density and mechanical property tests and a cosmic ray detector. Over its 322 Earth days of operations, it travelled 6.5 miles, returned more than 20,000 TV images, and performed a series of twenty-five X-ray fluorescence soil analyses at 500 different locations.

Lunokhod-2 (carried on the Luna 21 spacecraft), a more advanced robot, landed on January 15, 1973. Equipped with three slow-scan TV cameras mounted high on its rover for navigation, it returned images to ground controllers on Earth who sent real-time driving commands. Scientific instruments included a soil mechanics tester, solar X-ray equipment, and a French-supplied photodetector for laser detection experiments. It returned about 80,000 pictures over five months of operations.

Major achievements of Russia's Moon program also included three robotic sample return missions: Luna-16 (September 1970), Luna-20 (February 1972), and Luna-24 (August 1976).

All together they collected and returned 0.7 pounds (0.326 kg) of lunar surface materials, while, for comparison, the Apollo missions which returned a total of 842 pounds (382 kg).

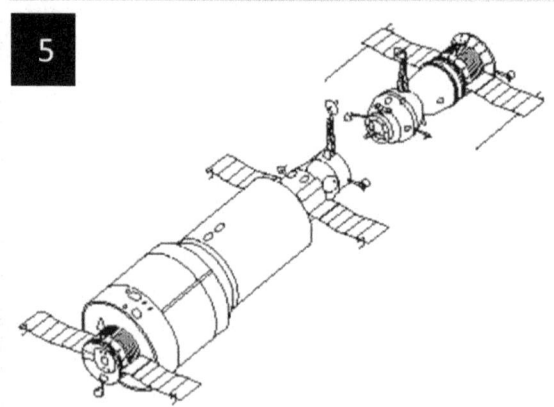

1. Neil Armstrong
   Source: NASA
2. Buzz Aldrin on the Moon
   Source: NASA
3. Soviet heavy lift Proton-K rocket
   Source: Astronautix
4. Space Shuttle Enterprise
   Source: NASA
5. Soviet Salyut-1 Space Station in
   1971.
   Source: Space.com

1. Apollo-Soyuz test project display at the National Air & Space Museum. Source: Joyofmuseums

2. Soviet Zvezda Module from the cancelled Mir-2 became the third module of the ISS in July 2000. Source: NASA

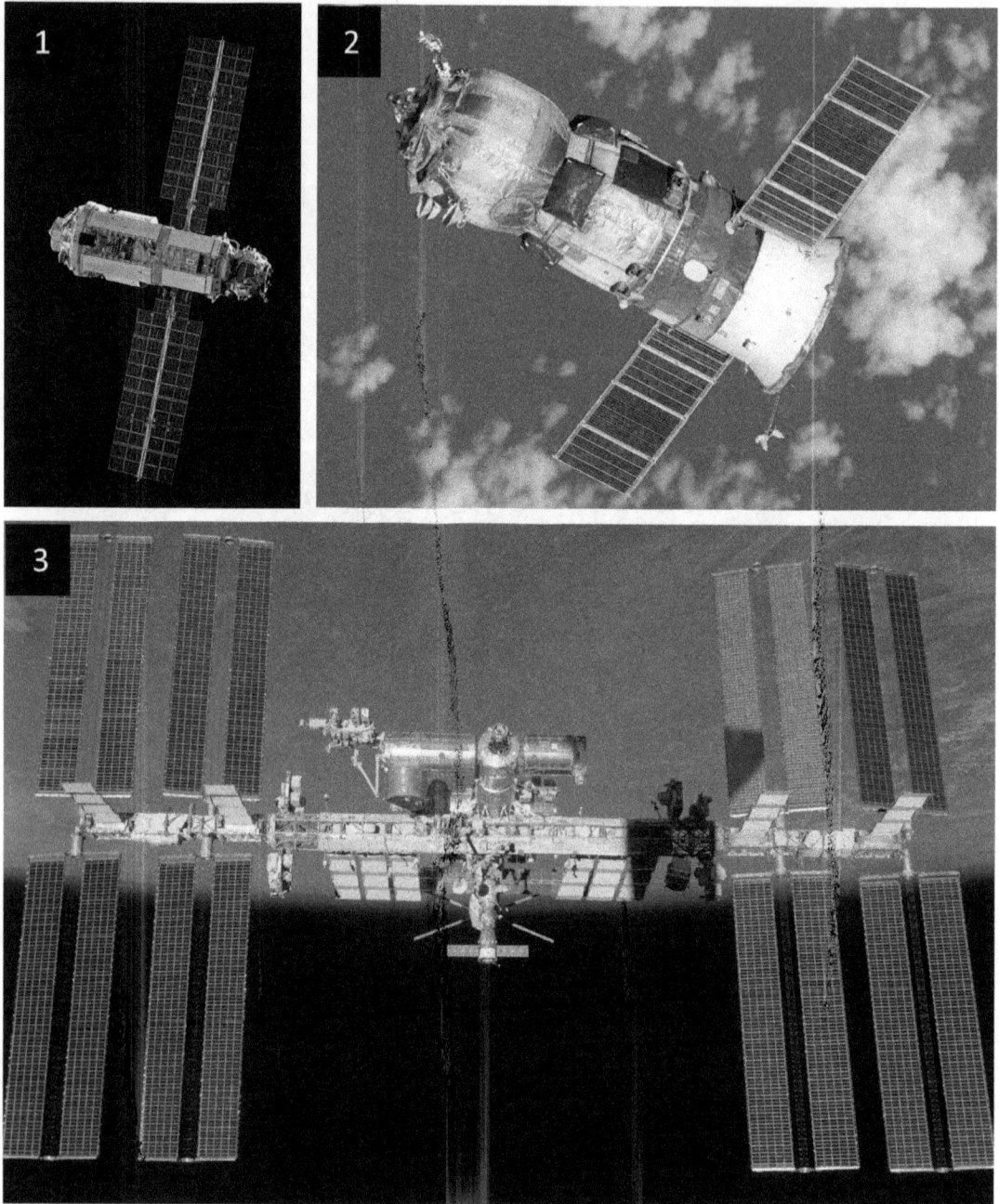

1. Soviet Zarya was the first module of the ISS lunched in 1998.
   Source: NASA
2. Progress Spacecraft
   Source: NASA
3. International Space Station
   Source: NASA

1. U.S. Skylab Space Station
    launched in 1973.
    Source: NASA

2. Soviet Mir Space Station
    assembled 1986 to 1995.
    Source: NASA

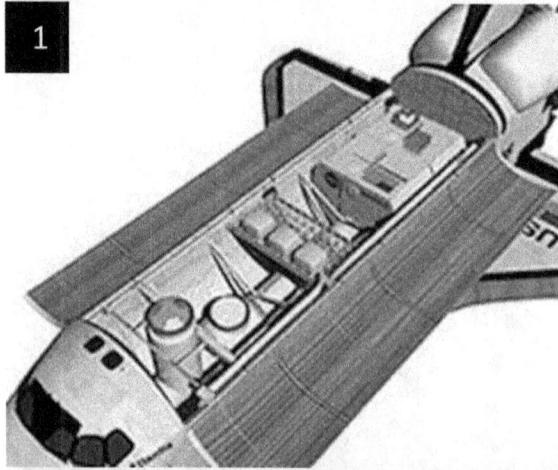

1. Spacehab
   Source: NASA
2. Scott Kelly
   Source: NASA
3. Space Shuttle Columbia
   Source: NASA
4. Cosmonaut Mikhail Kornienko
   served aboard ISS for one year
   in 2014. Source: NASA

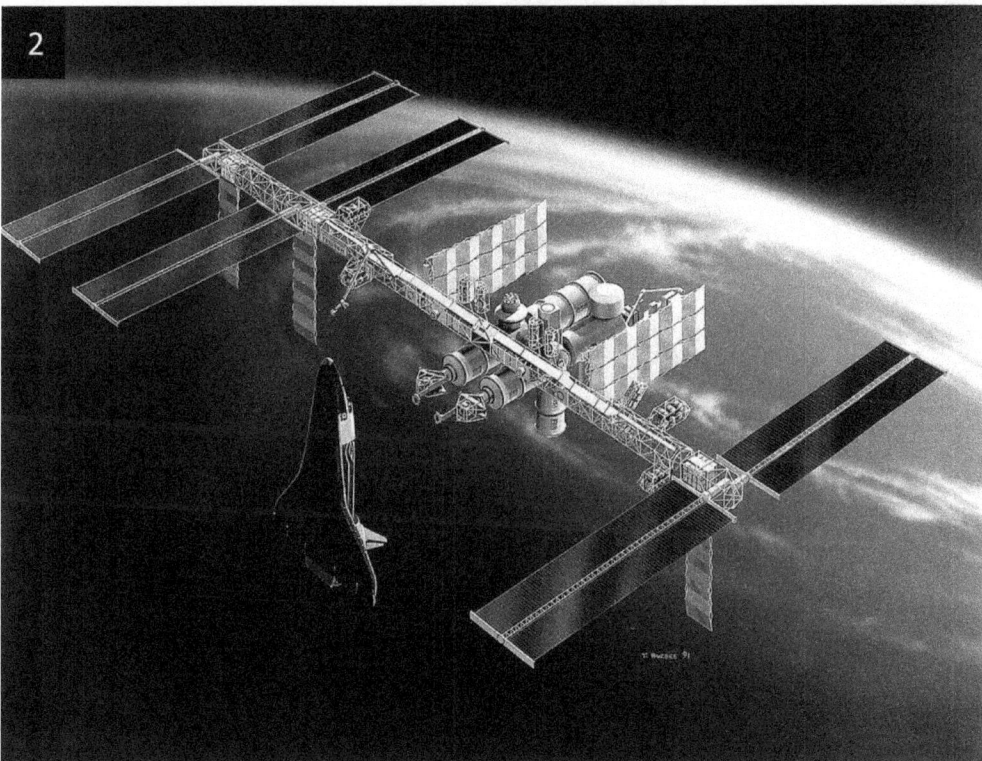

1. Privately financed *Industrial Space Facility* never flew into space. Source: Author

2. Space Station Freedom Source: NASA / Tom Buzbee

1. Titan IV Launch Vehicle
   Source: NASA
2. Space Shuttle Challenger
   Source: NASA
3. Soviet Buran Space Vehicle
   Source: RSC Energia
4. Space Shuttle Endeavour
   Source: NASA
5. Space Shuttle Discovery
   Source: NASA

1. J-2 Engine of the Saturn V S-IVB third stage. Source: Torsten Bolten

2. Space Shuttle Main Engine (R-25). Source: NASA

1. Christa McAuliffe, teacher killed in the Space Shuttle Challenger disaster. Source: NASA

2. Dennis Tito Source: NASA

1. Richard Nixon
   Source: Public Domain
2. Ronald Reagan
   Source: Public Domain
3. Mikhail Gorbachev
   Source: White House
   Photographic Collection
4. Larry Bell
   Source: Author
5. Guillermo Trotti, co-founder,
   Space Industries Inc.
   Source: SICSA
6. Maxime Faget
   Source: NASA
7. Jeff Bezos
   Source: Seattle City Council
8. Sir Richard Branson
   Source: Chatham House
9. Elon Musk
   Source: Duncan Hull

1. Johnson Space Center
   Source: NASA
2. Marshall Space Flight Center
   Source: NASA
3. Kennedy Space Center
   Source: NASA

1. New Shepard Rocket
   Source: ThePenulltimateOne
2. Falcon 9 Rocket
   Source: NASA

3. Crew Dragon Spacecraft
   Source: NASA

Chapter Eight: International Pathways Converge

THE 1970S AND 80s ushered in a new era of international space cooperation. With Apollo ended, priority Russian and U.S. attention shifted from lunar exploration to Earth orbital studies of human adaptation and mitigations associated with extended weightlessness, influences of weightlessness and space vacuum upon materials and physical/mechanical processes, and ways to enhance human safety and performance during future long-term multi-year missions.

Many nations accepted invitations to join and invest in common, peaceful enterprises of discovery in orbit and beyond. As a result, the world continues to witness benefits of multicultural engagement rising far above limited boundaries of interest defined on surface maps. This ongoing quest of discovery may ultimately prove to be the greatest and most rewarding space exploration challenge of all.

Global audiences watched on July 17, 1975 as TV images showed Soviet Cosmonauts Alexei Leonov and Valery Kubasov shake hands with NASA Astronauts Thomas Stafford, Vance Brand and Donald (Deke) Slayton high above the Atlantic Ocean. The historic docking of an American Apollo Command and Service Module (CSM) and Russian Soyuz vehicle mated critical mechanical pressure seals developed separately for the first time. This feat occurred after the two spacecrafts delivering them had made gradual trajectory changes over a two-day period after being launched within hours of each other. One had departed from the Kennedy Space Center in Florida...the other from the Baikonur Cosmodrome in Kazakhstan.

The political timing of the Apollo-Soyuz mission was no accident, highlighting a new policy of détente, a symbolic act of peace, between superpowers. The U.S. was engaged in a Viet Nam ground war which, given Russia's proxy involvement in the conflict, was adding to existing Cold War tensions.

The government-controlled Soviet press had been highly critical of America's Apollo program, printing on one occasion that:

> ...the armed intrusion of the United States and Saigon puppets into Laos is a shameless trampling underfoot of international law.

The article appeared above a photograph of an Apollo 14 launch seven months prior. This juxtaposition portraying American accomplishments in space with accusations of imperialist international agendas was part and parcel of Kremlin messaging.[62]

Soviet leader Leonid Brezhnev shifted that public position to extol peaceful diplomatic benefits of the Apollo-Soyuz experiment. He told the world:

> The Soviet and American spacemen will go up into space for the first major joint

*scientific experiment in the history of mankind. They know that from outer space
our planet looks even more beautiful. It is big enough for us to live peacefully on
it, but is too small to be threatened by nuclear war.*[63]

Apollo-Soyuz, the first joint U.S.-Russian space flight was to be the last flight for an Apollo spacecraft. It was also to be the last manned U.S. space mission until the first Space Shuttle launch in April 1981, providing important experience for future Shuttle-Mir and International Space Station (ISS) programs that followed. The Soyuz modules originally developed to carry cosmonauts to Salyut space stations and Mir are currently used for ISS. A minimum of two are docked there at all times to provide assured contingency departure for a crew of six (three passengers each).

The world's first space station, Salyut-1 launched in 1971, although not successfully inhabited, was originally developed by Korolev's design bureau under cover for a military space station secretly known as Almaz. The Russian ISS core module which provides all of the station's primary life support systems along with living quarters for two crew members appears to be based upon Almaz, which, in turn, appears to borrow design features taken from a small Manned Orbital Laboratory space station proposed by McDonnell Aircraft for the U.S. Air Force…a project that was cancelled. Cosmonauts from the Communist bloc and non-American astronauts from the West repeatedly set time-in-orbit endurance records aboard Salyut 4, 6 and 7 from April 1982 up through early 1991.

Many recognize Russia's establishment of their modules as the ISS core elements containing essential functions as a great and predetermined strategic coup. The Service Module, AKA, Zvezda (star) provides operational life support systems, while the Functional Cargo Block (FGB), AKA, Zarya (sunrise) provides electrical power, propulsion, guidance, and storage. Together, they put continuing dependence on all vital functions under their control, including periodic propulsive re-boosts as the ISS orbit decays over time—plus refueling which depends upon Russian Progress spacecraft.

Skylab, America's first space station, which orbited Earth from 1973 to 1979, was created using a converted third stage of a Saturn V Moon rocket outfitted with two decks as a habitat and orbital workshop. Strictly a U.S. NASA operation, the facility was spacious even by current standards, providing a large solar observatory and experiment area. A Crew Service Module (CSM), converted from the second stage of a smaller Saturn 1B booster, provided crew transport and emergency means to rapidly return to Earth.

Skylab's three total three-crew missions logged 513 man-days in orbit and accomplished thousands of experiments covering many different disciplines. Its orbit was allowed to slowly decay causing the facility to burn up on reentry five years after the last crew had returned home. Had a Space Shuttle then under development been available earlier, it might have re-boosted Skylab, extending its operational life by about five years. A Shuttle might also have provided precision control and thrust to guide Skylab to an assuredly safe and targeted final destructive reentry.

The Mir space station (which translated combines the words of world, peace and village) ultimately served as a true symbol of cooperation between peoples of Russia and the United States following a half-century of mutual antagonism.

Over its 15 years of operation—from the time its first module was launched on February 20, 1986—Mir hosted 125 cosmonauts and astronauts from 12 different nations who conducted approximately 23,000 scientific experiments. Ironically, on the very same day Mir's life ended on March 23, 2001, Russia expelled four U.S. diplomats and threatened to expel 46 more in retaliation for America's expulsion of 50 of theirs for suspected espionage.

America's Space Shuttles docked with Mir seven times. While prior to the Shuttle-Mir missions all experiments and supplies were provided exclusively by Russian "Progress" cargo vehicles, several Shuttle-Mir missions featured commercial services provided by SPACEHAB modules which are discussed later.

Unfortunately, the Progress ships provided no way to remove discarded items. By the time Americans first arrived about a decade into operations, American Astronaut Mike Foale described the cramped scene with broken equipment and floating bags of trash as "…a bit like a frat house, but more organized and better looked after."

As the world's first modular space station, Mir's outside appearance was cluttered too, variously characterized as a dragonfly with wings outstretched, a prickly hedgehog, and a 100-ton Tinker Toy. NASA-Mir Astronaut Jerry Linenger's description was more imaginative, appearing like:

> *...six school buses all hooked together. It was as if four of the buses were driven into a four-way intersection at the same time. They collided and became attached. All at right angles to each other, these four buses made up the four Mir science modules...Priroda and Spektr were relatively new additions...and looked it— each sporting shiny gold foil, bleached-white solar blankets, and unmarred thruster pods. Kvant-2 and Kristall...showed their age. Solar blankets were yellowed...and looked as drab as a Moscow winter and were pockmarked with raggedy holes, the result of losing battles with micrometeorite and debris strikes over the years.*[64]

Despite its inauspicious appearance, Mir served as a beautiful symbol of international cooperation in space science and collegiality. Sadly, for many, with a new International Space Station requiring much of the Russian space program's attention and resources, Mir ended its 86,000 total orbits as fragments in a watery grave. Vladimir Semyachkin—who had developed a variety of its control and navigation system—observed:

> *It's a shame...Our child, who we gave birth to so many years ago...we're going to have to put it to sleep. But, on the other hand, we understand that sometimes there's nothing to be done...One cannot sit, as it were, on two chairs at the same time. Nevertheless, despite this sorrow with...regard to Mir, we nonetheless do look forward to the future with a great deal of hope.*

The start of that future un-coincidentally coincided with the completion of the most complex international scientific and engineering project in history and the largest human structure ever to be put in space.

Planned and operated by five different agency partners representing 15 different countries, it serves a variety of purposes: a laboratory for biological, material, and other sciences; an observation platform for astronomical, environmental, and geological research; and a stepping-stone towards future space exploration. Responsibilities and investments are divided among NASA, Russia's Federal Space Agency Roscosmos, the European Space Agency, the Canadian Space Agency, and the Japan Aerospace Exploration Agency.

The scale of ISS is immense, spanning the area of a U.S. football field including end zones. Weighing a total of nearly a million pounds (450,000 kg) and orbiting at five miles per second (8,000 m/s) the complex now provides more livable space than a conventional five-bedroom house where a six-person expedition crew typically remains onboard from between four to six months. For comparison, ISS is nearly four times larger than was Mir, and about five times larger than Skylab.

During a former visitation with a docked Space Shuttle, the combined ISS complex has supported a total of thirteen people for several days.

The crew size was temporarily reduced to two-person teams after the tragic Columbia Shuttle disaster during which time the crew and supplies could only reach ISS using Russian Soyuz and Progress spacecraft. ISS hosted its first one-year crew in 2015-2016 involving NASA's Scott Kelly and Roscosmos's Mikhail Kornienko.

While ISS was originally planned to be decommissioned in 2020, international partners have yet to decide whether to grant a NASA request to postpone that schedule at least to 2024. A key obstacle to NASA's proposal centers upon deteriorated relations following Russian Military activities in the Ukraine. Just as the ISS demonstrates the substantial advantages that may be gained through international collaboration in space, this circumstance also demonstrates the serious risks attached to all large-scale international space programs.

Shuttle-Mir Commander Charlie Precourt predicted what others can still only hope, namely that the ISS would:

> *...provide the psychological impetus for politicians to force themselves to find an agreement to disputes that otherwise they wouldn't—because they'll all look up there and say, 'Well, we all have an investment in that too. We'll have to keep this relationship going in a proper direction.'* [65]

The international aspects of Russian ISS experience reflect U.S. vulnerability in depending upon other nations for development and operations of critical programs demanding that things happen when and how we wish them to. Continuing to do so will leave us exactly where we are, yet without partners NASA might never have had the resources to accomplish something like ISS.

This illustrates a conundrum attached to all massive international space programs.

Chapter Nine: Space Shuttle Controversies

COMPETING MILITARY AND civil interests plagued launch system priorities beginning from the time plans first emerged to support human and cargo space station operations. As a result, the Shuttle design and operation became a compromise among technical and political considerations…and between NASA and Department of Defense (DOD) priorities.

The DOD and the National Reconnaissance Office wanted a capability to launch 40,000-pound (18,150 kg) payloads to high-inclination polar orbits. They also preferred launching from the Vandenberg Air Force Base in California, the only site affording an ascent path to the south that wouldn't jeopardize safety of highly populated areas. The Space Shuttle couldn't meet that launch capacity at highly inclined orbits, and space station plans favored a lower-inclination twenty-eight-degree orbit as was initially planned for Space Station Freedom.

Frequent Shuttle launch delays and long payload manifesting preparations also created big problems for national security payloads. DOD realized that even if NASA was eventually able to achieve hoped-for 24 Shuttle launches per year, it still wouldn't accommodate Air Force needs in combination with civilian/private sector demands. In any case, the DOD went ahead and funded development of a Shuttle launch pad and preparation facilities at Vandenberg. At the same time, the Air Force also pushed for approval of what they called a Complementary Expendable Launch Vehicle (CELV) to provide assured access to space. The term complementary was clearly intended to minimize the appearance of competition with the Space Shuttle.

Congressional reaction was mixed because opponents knew that the CELV would divert DOD use away from the Shuttle and undercut its funding. Nevertheless, the CELV—which became known as the Titan IV—was approved and first launched on June 14, 1986.

A chronic inability of the Shuttle to meet DOD requirements combined with the loss of Challenger and development of the Titan IV dramatically impacted commercial launch markets.

The Challenger tragedy prompted the Reagan administration to reconsider what types of payloads Shuttle should carry, ultimately restricting them to its unique capability requirements and national security purposes. Commercial satellite launches were excluded, and other private market uses were forced to compete all the more with DOD priorities. It didn't help that alternative American launch capabilities were allowed to deteriorate in response to the planned omnipotence of the Shuttle as a launch system. As a result, U.S. launch expendables had to rebound to pick up the slack.

Meanwhile, DOD, upon having reconsidered Shuttle risks for critical payloads as well and shifting to Titan IV dependence, abandoned plans to use the Vandenberg Shuttle launch site altogether after already spending several billions of dollars upgrading it.

There's less reason to wonder then why, after spending all that money, the Russians saw the U.S. Shuttle as a likely weapons delivery system…a belief that led them to create their own Buran version as a counter defense.

Chapter Ten: Grounded!

RATHER THAN DISPUTE whether or not the Space Shuttle was taken out of service prematurely, let's review some background events leading to that decision.

Unlike Kennedy's May 25, 1961 we-should-go-to-the-Moon speech, the Space Shuttle program gradually emerged over time, beginning when NASA had its plate full of responsibilities to complete work on Saturn V and get Apollo astronauts safely to the Moon and back. As the agency began to think about maintaining manned programs beyond Apollo, the top priority became gaining congressional approval for an orbital space laboratory—a space station.

Yet since doing so would require affordable and reliable means to transport crews and supplies back and forth, emphasis upon the most immediate need shifted to a focus creating an Earth-to-orbit transportation system—a Space Shuttle.

NASA convened a Space Shuttle Task Group in December 1968 to determine what basic missions and vehicle characteristics should guide development. The group concluded that the orbital transportation system should provide space station logistical support; should enable the launch and retrieval of propulsive stages, payloads and satellites; should support satellite servicing and maintenance; should accommodate propellant delivery; and should enable short duration manned orbital missions.

Economy and safety for what was then termed an Integral Launch and Reentry Vehicle (ILRV) were to take precedence over optimized payload capacity, and contracts to investigate preliminary concept options were awarded to General Dynamics, Lockheed, McDonnell Douglas and North American Rockwell.[66]

In response to a directive by President Richard Nixon calling for a high-level study to recommend a future course of action of the overall civilian space program, a Special Task Group (STG) chaired by Vice President Spiro Agnew delivered a September 15, 1969 report recommending that a fully reusable shuttle concept be developed to operate in an airline-type mode.

The premise was that a Reusable launch Vehicle (RLV) would:

> *…provide a major improvement over the present way of doing business in terms of cost and operational capability.*

It would:

> *Carry passengers, supplies, rocket fuel, other spacecraft, equipment, or additional rocket stages to and from Low Earth Orbit (LEO) on a routine, aircraft-like basis.*

The new RLV would:

> *Be directed toward supporting a broad spectrum of both DOD and NASA missions.*

As the STG report envisioned, the space transportation system would really be comprised of three different elements. One was a chemically fueled shuttle repeatedly operating between Earth and LEO. The second was a chemically fueled space tug or vehicle for moving people and equipment to different Earth orbits and as a post-Apollo transfer vehicle between a lunar-orbit base and the lunar surface. Finally, a reusable nuclear stage would transfer people, spacecraft and supplies between Earth orbit, lunar orbit and geosynchronous orbit (where satellites can retain fixed positions over points on the Earth surface), and as needed for other deep space activities. The plan envisioned all three capabilities collaboratively operating in unison to support a comprehensive set of capabilities. Of these three, only the first—which no longer operates—was ever accomplished.

Whereas logistical functions in support of a proposed space station in Low Earth Orbit (LEO) had originally been cited as the principal Space Shuttle justification, it soon became apparent to NASA that neither the Nixon administration nor Congress would endorse simultaneous development of both programs. This realization first led NASA to refocus attention upon creating a two-stage, fully reusable Shuttle. Further investigation of costs and technical difficulties in accomplishing a full-fly-back capability of the enormous first booster stage and the operational vehicle prompted a preferred hybrid approach where the manned truck-like vehicle returned separately. NASA and aviation industry engineers ultimately rejected that approach based upon a lack of technology then available to sustain extremely high heat loads along with generally poor knowledge about atmospheric reentry of large structures at the time.

It was also determined that using existing F-1 and J-2 first stage engine technologies, both of which were already out of production, would be inadequate to meet safety and weight requirements for Shuttle application without significant redesign. Accordingly, in July 1971, NASA awarded a development contract to Rocketdyne to develop a completely new engine which came to be known as the Space Shuttle Main Engine (SSME).

By late 1971, designers within NASA and industry began to converge on a vertically launched delta-winged Shuttle vehicle concept which incorporated a low-thrust orbital maneuvering capability. The orbiter spacecraft would be mounted to a large, releasable external tank carrying liquid oxygen and liquid hydrogen propellant. The external tank would be flanked by expendable strap on booster rockets which were originally intended to be liquid fueled. NASA later opted for solid-fuel technology. Putting all launch fuel and oxidizer in detachable tanks simplified orbiter design and allowed the Shuttle orbiter to carry a greater payload as a percentage of total weight than a two-stage fully reusable system. It also simplified the design and propellant tank construction, thereby reducing cost.

Official Nixon White House approval of the Space Shuttle was announced on January 5, 1972…a decision very likely influenced in part by positive presidential election benefits to be gained that year by a large aerospace program, most particularly with regard to significant employment impacts in key electoral states.

The plan to use the solid fuel and oxidizer—the Solid Rocket Booster (SRB)—as the first stage strap-on won out over liquid chemicals by reasons of lighter weight combined with simplicity, reliability and easier ability to be refurbished after being recovered from corrosive ocean effects. A primary SRB disadvantage is a lack of means to shut the engines down once ignited. Another disadvantage involved the SRB's substantially higher acoustic and vibratory loads. Both problems, however, were offset by lower initial and recurring costs.

A contract to design and develop the Shuttle orbiter was awarded to North American Rockwell under supervision of the Lyndon B. Johnson Manned Spacecraft Center in Houston. Morton Thiokol was selected to produce the SRB under supervision of the Marshall Space Flight Center in Huntsville, Alabama which was also responsible for developing and manufacturing the Space Shuttle Main Engine. The Kennedy Space Center was put in charge of developing methods for Shuttle assembly, checkout, and launch operations.

The challenges overcome in bringing the Space Shuttle into operation were no small achievement. The United States as of that time had never built a new SSME rocket engine that was both reusable and capable of being throttled. This required development and demonstration of high-pressure turbopumps capable of higher speeds and internal pressures than previously achieved.

Reliable thermal protection for the Shuttle orbiter during Earth reentry presented another major problem. Whereas Mercury, Gemini, and Apollo capsules accomplished this using ablative materials that heated up and burned off upon encountering the upper atmosphere, these structures were used only once. Shuttle orbiters, on the other hand, would need a solution robust enough to sustain numerous launches and returns, while also being light enough to keep the spacecraft's weight within stringent limits.

This was ultimately achieved by applying more than thirty thousand low-density ceramic heat shields (tiles) bonded to a lightweight aluminum attachment structure that would deform slightly under aerodynamic loads. Since individual tiles varied in size and shape, each carried a special identification number and installation code. An intermediary bonded layer of compliant materials, similar to that which covers tennis balls, isolated the tiles from the airframe. Surfaces of the Shuttle not directly subject to reentry heating were sheathed in a heat-resistant beta cloth material.

As it turned out, the Space Shuttle never proved to be as economical as hoped, and as Challenger and Columbia tragedies demonstrated, were not the routinely reliable and economical airplane-mode space trucks much of the public came to expect either. The second point will be discussed in more detail later.

Houston, We Have a Problem

A catchphrase reference to Apollo 13's perilous 1970 mission now applies to the U.S. space program overall… "Houston, we have a problem." Adrift in a cosmic vacuum of public apathy, NASA's present state of organizational inertia reflects an absence of national leadership with bold vision, clear goals, and serious commitments. This previous erosion of general citizen interest began soon after presidential and congressional politics turned the Apollo space program into a jobs program.

The sad truth is that while most of us take great pride in the amazing things NASA has accomplished, few, even within the agency, know where that legacy is headed. Although NASA's key International Space Station role represents an epic achievement, any long-term follow-on plans, either in low-Earth orbit or beyond, are obscure. And while the Space Shuttle was a marvelous invention, is now being proven that commercial enterprises can build and operate future space transportation vehicles far more economically.

Giving NASA and its contractors due credit, let's remember that the Space Shuttle Columbia launched astronauts John Young and Robert Crippen to orbit in 1981, just nine years after the White House approved the program, successfully doing so with no prior unmanned tests. The main engines of that entirely new and very complex vehicle were each about the size of a passenger car yet produced more power than all engines of the Titanic, which covered an area the size of about 15 tennis courts.

Ironically, the tragic 2003 loss of that same Columbia spacecraft and crew following the deadly 1986 Challenger disaster prompted a decision to wind down and mothball the Shuttle program. That early chapter of space development also turned out to be more expensive than was originally estimated. Whereas each launch was projected to cost $110 million, the actual figure came much closer to $1 billion. By comparison, Elon Musk's SpaceX now convincingly intends to set future commercial launch prices at considerably less cost.[67]

Chapter Eleven: NASA Flirts with Space Commercialization

EARLY COST/BENEFIT ESTIMATES foresaw Space Shuttle flight rates progressively growing to forty-eight annually by the mid-1980s, with each of the several fleet vehicles operating at short turn-around schedules.

These projections were based upon unrealistically overestimated revenue-producing international government, civilian commercial and military utilization and unrealistically underestimated time and cost requirements to refurbish returned vehicles for next-flight opportunities. Yet, despite a Ronald Reagan administration policy encouraged by NASA to phase out Delta, Atlas, and Titan Expendable Launch Vehicle Systems (ELVs) which competed for commercial satellite launch markets after 1982, annual Shuttle launches had increased only to nine by 1985. The European-built and largely government-subsidized Ariane rocket afforded the only comparably capable alternative for launching commercial payloads.

Rising dissatisfaction among commercial groups over what was amounting to a twenty-four-month period required to prepare payloads for launch aboard the Shuttle, coupled with high costs, motivated the Reagan White House to a policy shift that would put Shuttle operations under control of the Air Force or private sector.

Presenting similar concerns, the Department of Defense argued that long Shuttle manifest requirements and frequent launch delays were unacceptable for time-critical national security payloads, and that a catastrophic system failure would ground their operations altogether.

In addition, the Shuttle proved incapable of delivering 40,000-pound (18,100 kg) payloads to polar orbits as required for some strategic satellite deployments of that era. A Shuttle launch facility originally intended to host Shuttle launches to polar orbits was built at Vandenberg Air Force Base in California, but it was never used.

The DOD countered with a proposed supplementary Complementary Expendable Launch Vehicle (CELV) of its own, to provide assured access to space. That program, later known as Titan IV, was approved, with a capacity to deliver 40,000 pounds (18,100 kg) to a high-inclination low Earth orbit, and 10,000 pounds (4,535 kg) to geostationary orbit. This was accomplished by stretching the original Titan liquid propellant tanks and upgrading its strap-on solid rocket motors to seven segments rather than the five and a half used by the Titan 34D.

With Titan IV usurping the DOD military markets NASA had counted on and Ariane competing for international and commercial satellite launch business, NASA looked to a brand-new revenue opportunity with an added potential for marketing a previously deferred space station initiative. Following completion of the transportation system that would provide access to LEO, then-NASA Administrator James M. Beggs called for the next logical step, the U.S. answer to Mir.

President Reagan announced support for the plan in his 1984 State of the Union Address, stating:

> *America is too great for small dreams... We can follow our dreams to distant stars,*
> *living and working in space for peaceful economic and scientific gain.*

Dubbed Space Station Freedom, the facility was to function as an orbiting satellite repair shop, an assembly platform for spacecraft, an observation post for astronomy, and most prominently advertised, as a laboratory for commercial users interested in studying effects of weightlessness on physical processes and materials.

But, just to be clear, the term weightlessness doesn't really mean that there is an absence of gravity, but rather that there is an absence of relative motion between objects in a free-falling environment. This might be imagined by visualizing someone on an elevator dropping a coin. That coin will drop to the floor so long as the elevator and everything in it is descending at a uniform speed. On the other hand, consider what would happen if the cable breaks so that everything, the elevator, he/she and the coin are in free fall together. In that unfortunate event for the passenger, the coin would literally float relative to the surrounding elevator until all come to an abrupt stop.

Weightless (or more technically accurate microgravity) effects occur when a spacecraft's orbiting speed exactly counters the effects of gravity. Without gravity, the spacecraft would shoot off into space like a stone from a slingshot. Such conditions afford opportunities to accomplish high-value scientific studies not possible on Earth…such as processing of new and improved chemical compounds for use in medicine and industry.

Prospective space processing markets are numerous and varied. Examples include: growth of advanced crystals of compounds and alloys used to improve semiconductor technology for the electronics industry; container-less processing of special glasses while positioned in a furnace by acoustic pressure to understand how to improve glass materials for optical and electrical applications; new alloys made possible by avoiding separations of metals of different density (like trying to mix oil and water), leading to unique structural, electrical and magnetic materials; and achieving improved separations of purer biological samples which enable pharmaceutical companies to expedite FDA approvals for medical uses.

In the late 1980s, NASA prominently featured space material processing and revenue benefits in an expansive Space Station Freedom marketing campaign aimed at winning congressional, public, and commercial industry support. Upon reviewing potentials, a start-up company co-founded by editor Larry Bell, Guillermo Trotti, James Calaway, and former NASA Chief Engineer Maxime (Max) Faget responded with a plan to incubate and develop commercial processing opportunities. As proposed, Space Industries, Inc. (SII) would privately finance, develop, and operate an Industrial Space Facility (ISF) which would co-orbit with Freedom.

Scheduled for launch in the early 1990s, the ISF was designed as a free-flying, pressurized laboratory with 18 kW of power to accommodate plug-in microgravity equipment. Lacking independent life support systems, it was to be occupied only when a Shuttle orbiter was docked with it during equipment servicing and change-out operations. A major ISF advantage was to provide extended-length experiment opportunities under a very high-level microgravity environment required by some experiments.[68]

SII raised several million dollars of private investment to undertake ISF planning and design, supported by Westinghouse and Boeing as an in-kind service contributing partners. NASA originally agreed to provide two-and-one-half dedicated Shuttle launch and servicing missions in exchange for partial anchor tenant use of the facility through a formal fly now-pay later arrangement.

The company's plans and promising prospects abruptly changed following the posting of a January 15, 1988 front page New York Times feature titled *A Rival for the Space Palace* that presented the ISF as a cheaper and preferable alternative to NASA's planned space station rather than as a complement as SII intended.

As the damaging article stated:

> *The space station NASA wants to build is a luxury sky hotel with a sky-high price tag, now at $32 billion. Congress has at last decreed that the agency should also consider supporting a much smaller, privately operated space station that costs a mere $700 million and can do many of the same things. NASA seems petrified that Congress may next wonder why it needs a palace in space if a mobile home would do nearly as well.*

Critical of NASA's budgetary priorities, the article went on to say:

> *The lab thus seems a better bet than investing in a full-fledged space station right away. Yet NASA doesn't see it that way. The space agency apparently prefers hardware to results. If cheap access to space had been the agency's top priority, it could have pressed long ago to reduce the cost of launching payloads from the present $3,600 per pound to $400 per pound, a project the Air Force and NASA have just begun.*

The article also cited ISF benefits to the Space Shuttle:

> *The agency says it has no need to lease space on the private orbiting lab, and that the lab does not compete with the space station: most experiments need to be continuously watched, as the Space Station makes possible, not merely visited every four months. But the orbiting lab, with its big solar panels, could supply the power to double the Shuttle's time in space, allowing for some extended experiment watching. And doubtless some experiments now designed for continuous monitoring could be adapted to the lab.*

NASA then decided that it needed to restudy its original agreement with SII and consider other commercial lab options, ultimately electing to pursue a pressurized research and cargo module that would be carried up and down during dedicated research missions in the Space Shuttle payload bay. The contract was awarded to SPACEHAB, Inc. who later leased the modules to NASA on a mission-by-mission basis.

A tunnel connected the pressurized SPACEHAB to the orbiter's crew compartment, so that astronauts could reach it—and work inside it—without putting on spacesuits. On some missions, two modules were flown joined together to provide additional room for experiments and storage.

Commencing with a June 1993 flight, SPACEHAB modules, and later also unpressurized cargo carriers, flew on 22 Space Shuttle missions, including eight resupply missions to the ISS and seven to Mir. The tragic loss of Space Shuttle Columbia, which was carrying an only partially insured SPACEHAB new science double-module, was a major financial setback. However, SPACEHAB continued to support ISS assembly and logistics missions using its remaining pressurized modules and unpressurized pallets, the latter becoming permanent ISS installations.

SPACEHAB's pressurized module design was patterned after a similar Shuttle payload lab module program called Spacelab developed in the 1970s and early 1980s by the European Space Agency (ESA) in exchange for flight opportunities. The system included two types of elements: pressurized modules operated from inside the payload bay, and unpressurized pallets which were exposed directly to space. Spacelab directly competed for SPACEHAB markets for missions designed to accommodate science operations.

Twenty-two Spacelab missions flew until the program was retired in 1998. The first full mission took place aboard Space Shuttle Columbia on November 28-December 8, 1983. They were interrupted along with all other manned activities following the loss of Space Shuttle Challenger on January 28, 1986 until 1988 when flights resumed.

NASA's early resistance to commercial space capabilities reflected a perceived threat to their own interests and control.

This perception was made abundantly clear in the agency conflicts with Russia over the advent of what would become known as space tourism. MirCorp, a U.S. commercial space company, engaged the Russian Space Agency in a plan to restore operations of the aging Mir space station under commercial management as a low-cost alternative to the ISS. In doing so, MirCorp consummated a lease agreement with the Russian space company NPO Energia in February 2000 which owned commercial rights to the space station to use the facility as a media entertainment and space research platform. The deal provided NPO Energia with a 60 percent majority ownership.

MirCorp's financial backing was designed to enable NPO Energia to re-boost Mir, and thereby postpone a deorbit that had previously been agreed to by the Russian Space Agency in discussions with NASA and in anticipation of Russia's involvement as a partner in the ISS program. Without MirCorp's commercial investment, Russia lacked the money to save Mir, but could reallocate its resources toward its new ISS obligations.

In addition, MirCorp financed the first privately funded manned expedition to a space station (Soyuz TM-30 from April 4-June 16, 2000); the first privately funded resupply mission in space (April 12, 2000); the first privately funded spacewalk (May 12, 2000); and the first space tourist contract (Dennis Tito, June 19, 2000).

None of these developments were welcomed at all by NASA, whose officials criticized the arrangements for interfering with international space policies. Nevertheless, the April 4, 2000 mission carried two Russian crew members, Sergey Zalyotin and Aleksandr Kaleri, to Mir where they undertook 73 days of refurbishment including locating and repairing a fluid leak.

A June 2000 press conference held at the Russian Mission Control Center announcing MirCorp plans to fly wealthy former U.S. space program engineer and investment tycoon Dennis Tito as the first Citizen Explorer aboard the Russian ISS module raised enormous controversy within NASA and Congress. NASA reportedly responded by cancelling planned high-level meetings with Russian Space Agency officials and declaring that a bill would be sent to them for any damages that might result.

Despite NASA's objections, Dennis Tito's flight aboard ISS proceeded. On April 28, 2001 he joined the Soyuz TM-32 mission to spend 7 days, 22 hours orbiting Earth 128 times to become the first bona fide space tourist. He reportedly spent $20 million for the ticket.

Other tourists followed Tito's earlier path to the ISS aboard Russian spacecraft, albeit at increasingly high ticket prices. Eventually Russia's Soyuz seats on ISS-bound missions gained value after NASA lost its own ability to transport its own crews aboard the Shuttle.

The Tito tourism venture might be credited as the business model leading to plans by Virgin Galactic founder Sir Richard Branson to market suborbital flights and his July 10, 2021 launch, along with six crew and passengers, aboard his SpaceShipTwo vehicle.

Similarly, Amazon founder billionaire Jeff Bezos along with three others traveled to the edge of space on an 11-minute flight aboard his com Blue Origin company's New Shepard rocket just over a week later on July 21.

NASA's resistance to commercial competition appeared to lessen with the appointment of NASA Administrator Michael Griffin in 2005. During a speech soon after taking charge, he said:

> Those of us on the government side of the space business must recognize a fundamental truth: if our experiment in expanding human presence beyond the Earth is to be sustainable in the long run, it must ultimately yield profitable results, or there must be a profit to be made by supplying those who explore to fulfill other objectives.

Griffin concluded…

> We should reach out to those individuals and companies who share our interest in space exploration and are willing to take risks to spur its development.[69]

Although this change of heart came too late for Space Industries, Inc. and SPACEHAB, it did usher in the current era of collaboration between NASA and commercial transportation service providers.

Today, NASA utilizes commercial delivery for ISS cargo missions, and is already purchasing commercial transportation for ISS crews and cargo from Elon Musk's SpaceX aboard the company's reusable Falcon 9 rocket and Crew Dragon capsule.

Chapter Twelve: Losses of Innocence and Invincibility

WITH APPROPRIATELY DESERVING recognition of those visionary space entrepreneurs and adventurers who will expand boundaries of human progress, it is naïve to underestimate attendant risks and costs. Numbing January 28, 1986 TV images of Space Shuttle Challenger—with seven crew members aboard—disappearing behind a huge fireball just over a minute after liftoff dispelled complacent views of human space flight as a routine, casual endeavor. Seventeen years later, as Columbia disintegrated upon reentry, another tragic loss of seven others underscored that unfortunate and inescapable reality.

Notwithstanding the Space Shuttle representing an indisputably marvelous engineering achievement, harsh operational strains and inevitable technological obsolescence conspired against the Space Shuttle's reliable and practical life. As particularly true in the case of all human launch systems, safety is of utmost concern. The Shuttle's reusable nature and versatile flight and aircraft-mode landing characteristics imposed additional complexities and vulnerabilities. Further, the high cost of assimilating rapid technological advances into a vehicle that demanded complex certification and integration resulted in a fleet embodied with increasing obsolescence and upgrade limitations.

On February 3, 1986, President Reagan charged a high-level commission chaired by former Secretary of State William P. Rogers with responsibility not only to find out what technical failure caused Challenger's catastrophic failure, but also to determine why such an oversight had been allowed to happen. Although senior NASA officials had preferred to carry out their own investigation outside the glare of publicity as they had done following the Apollo fire, the dramatic loss of life made an independent external review unavoidable. After all, NASA had promoted even more public interest in the flight by featuring crew member Christa McAuliffe as the first teacher in space, reflecting its public position that the Shuttle was a fully operational vehicle...not one which was still subject to substantial risk and uncertainty.

The newly formed Presidential Commission on the Space Shuttle Challenger Accident enlisted investigatory help from other federal agencies and private experts. Included were corporate experts who were familiar with the Shuttle's many systems and points of weakness, veterans who were intimately familiar with NASA and contractor cultures, and seasoned personnel with detailed understanding about critical safety processes and protocols.

The open Rogers Commission hearings and report, which NASA officials had opposed, provided extraordinary insights into nearly overwhelming complexities of preparing and approving the Shuttle for flight. The technical failure was traced not only to an O-ring seal malfunction in connecting sections of Solid Rocket Boosters (SRBs), but also to a failure on the part of launch officials to recognize excessively cold weather conditions jeopardized the performance of those joints. It was revealed that although several NASA and Morton Thiokol engineers were aware of a design deficiency, both organizations neglected to redesign the joint, resorting to other fixes—including tightening the fit and adding putty to assist the O-ring seal.[70]

The hearings publicly exposed a large number of crucial management deficiencies within NASA, including communication and coordination difficulties between mid-level NASA engineers and contractor personnel in conveying known problems to senior-level managers. It also became evident that NASA officials had subtly, but inexorably, shifted attitudes regarding launches to positions requiring that engineers must demonstrate that the Shuttle was unsafe—rather than safe—to launch.

A particularly dramatic moment at the hearing occurred when commission member Richard Feynman placed a short O-ring section in ice water, demonstrating on live television how inflexible the material became. As Morton Thiokol engineer Roger M. Boisjoly noted, decision-making regarding the Shuttle had become:

> *…a kind of Russian roulette… [the Shuttle] flies [with O-ring erosion] and nothing happens. Then it is suggested, therefore, that the risk is no longer so high for the next flights. We can lower our standards a little bit because we got away with it last time… You got away with it, but it shouldn't be done over and over like that.*[71]

During the hiatus before returning the Shuttle to flight status, NASA examined virtually every vulnerable element, fully rethought launch preparation and operations protocols, instituted many safety procedures, and replaced several system components. NASA also increased its contractor staff at the Kennedy Space Center, and more than doubled the time allocated to refurbish orbiters after flight for next launches.

These changes made it increasingly doubtful that a previously hoped-for annual twenty-four flight launch rate could be achieved even with more funding. This circumstance put an even bigger damper on Shuttle markets for commercial communications satellites which could be routinely and less expensively launched using expendable rockets. Europe's Ariane business benefited, servicing much of that market gap.

The loss of Challenger and its crew forced Reagan administration officials to reconsider policies regarding which types of payloads the Space Shuttle should carry and what sort of missions it would be used to support. Responding to long-standing criticism that it was a waste of federal resources to subsidize costly Shuttle launches of commercial satellites which could be delivered more cheaply by expendable rockets, a new policy restricted uses to those requiring its unique capabilities or which were attached to national security purposes. Still, anticipating a continuing demand for Shuttle services, The White House and Congress awarded Rockwell International a contract to build a Challenger orbiter replacement, the Space Shuttle Endeavor, which was first launched in May 1992.

The Space Shuttle system returned to service with the Discovery liftoff of STS-51L on September 29, 1988. An Aviation Week & Space Technology editorial observed:

> *The launch witnessed by the largest gathering of spectators and press since the Apollo 11 launch to the Moon in 1969, was the balm to the wounds remaining from the Challenger accident. It was a long time coming…it was a moment worth waiting for… The Discovery mission should be savored as a triumph for NASA, the US space program and the nation.*[72]

NASA had originally designed the Space Shuttle orbiter with an expectation that each vehicle in the fleet would have a 97% probability of lasting 100 flights, with each individual flight having a reliability of at least 99.97%. One NASA-funded study estimated that the reliability would be more like between 97-98%. Following the Challenger (STS-51L) disaster, and with little history of successes and failures to upon which to base such analyses, the congressional Office of Technology Assessment assumed for illustrative purposes that reliability would be 98%, and that NASA would face a fifty-fifty chance of losing an orbiter within thirty-four flights.[73]

Tragically, that OTS estimate came pretty close to being right. On February 1, 2003, Space Shuttle Columbia, STS-107, broke up during Earth reentry killing seven more astronauts including Israeli Space Agency payload specialist Ilan Ramon. The cause, a large piece of insulating foam on a bipod ramp attaching the external tank to the orbiter had fallen off during launch, fatally breaching the structural integrity of its left wing's leading edge.

An investigation board determined that a potential foam problem had been known for years. Making matters even worse, several people within NASA who saw the foam impact on video had unsuccessfully pushed to get pictures of possible wing damage during Columbia's 16 days in orbit. A Columbia Accident Investigation Board (CAIB) investigation concluded that although the Department of Defense had offered to use its orbital spy cameras to get a closer look, NASA officials in charge declined the offer.

Weeks after the disaster, CAIB released a very critical multi-volume report placed the blame for the Columbia disaster on a NASA culture that had minimized known safety problems over years.

The board found:

> *Cultural traits and organizational practices detrimental to safety were allowed to develop; that reliance on past success* [had been applied] *as a substitute for sound engineering practices; and noted organizational barriers that prevented effective communication of critical safety information.*[74]

These assessments virtually repeated previous NASA negligence assessments leveled by the Rogers Commission regarding causes leading up to the Challenger disaster.

CAIB not only recommended that NASA ruthlessly work to eliminate Shuttle safety problems but proposed that it be replaced by an entirely new space transportation alternative.

Its report stated:

> *The Shuttle is now an aging system, but still developmental in character. It is in the nation's interest to replace the Shuttle as soon as possible.*

Chapter Thirteen: Russian Shuttle as a Star Wars Defense

IN NOVEMBER OF 1988, during the regime of General Secretary Mikhail Gorbachev, the Soviets launched what appeared to be a virtual copy of America's Space Shuttle called Buran (meaning snowstorm). Closer examination, however, revealed significant differences.

First, it operated automatically and had no crewmembers aboard. Second, unlike the American Shuttle, the Buran orbiter was designed as a stand-alone vehicle which carried no rocket engines. Instead of integrated SRBs, an external tank and Space Shuttle Main Engines, it was carried to orbit aboard an all-liquid Energia heavy-lift launcher capable of delivering 220,000 pounds (100,000 kg) to orbit.

Additional differences involved propulsion capabilities. Whereas the American Shuttles had a maneuvering system that enabled orbital adjustments, the Buran did not. On the other hand, while Buran had jet engines to assist landing maneuvers, NASA's Shuttle landed as an unpowered glider.

Even more different than its U.S. counterpart, Buran (one of three produced) only flew once, and then a total of only two orbits. Although the flight appeared to be a highly successful achievement, none was ever repeated. Many wonder why.

James Hartford presents an interesting perspective on this issue in his book *Korolev: How One Man Masterminded the Soviet Drive to Beat America to the Moon* quoting insights taken from a discussion with former Soviet engineer Efraim Akim who had worked on the first robotic mission to succeed in making a soft landing on the Moon and send back photos.

As he explained:

> *When the US Shuttle was announced we started investigating the logic of that approach. Very early calculations showed that the cost figures being used by NASA were unrealistic. It would be better to use a series of expendable launch vehicles.*

Akim continued:

> *Then, when we learned of the decision to build a Shuttle launch facility at Vandenberg AFB for military purposes, we noted that trajectories from Vandenberg allowed an overflight of the main centers of the USSR on the first orbit. So, our hypothesis was that the development of the Shuttle was mainly for military purposes. Because of our suspicion and distrust, we decided to replicate the Shuttle without a full understanding of its mission. When we analyzed the trajectories from Vandenberg we saw it was possible for any military payload to*

reenter from orbit in three and a half minutes to the main missile centers of the USSR, a much shorter time than an SLBM [Submarine-Launched Ballistic Missile] could make possible (ten minutes from the coast).

Akim concluded:

You might feel that this is ridiculous, but you must understand how our leadership, provided with that kind of information, would react. Scientists have a different psychology than military. The military is very sensitive to the variety of possible means of delivering the first strike, suspecting that the first strike capability might be the Vandenberg Shuttle's objective, and knowing that a first strike would be decisive in a war, responded predictably. That's why confrontation is so dangerous. Fortunately, it's all academic now. Slava bogu [thank God].[75]

In his book, James Hartford also presents a corroborative discussion with Boris Gubanov, a key figure in the Buran development. Gubanov emphasized that the decision to build Buran as an established program was further stimulated by congressional approval of President Ronald Reagan's plan to develop a Strategic Defense Initiative (SDI) where America's Shuttle was feared as a first-strike hydrogen bomb delivery threat.

Gubanov is quoted to say:

When the SDI program emerged, it was described as a high-energy launch vehicle, capable of reacting to communications satellites and strike weapons, that would be in orbit for a long time. The Space Shuttle should have the capability of rendezvousing with what would be a 20-ton SDI vehicle and returning it safely to Earth. Nobody had yet developed such a system. That was what Buran and the US Shuttle were supposed to do.

Gubanov added:

Now that we have peace, we have to ask why we have these systems.[76]

The short answer is that we don't...not at the required price. The Buran was mothballed in 1993 after an SDI-assisted Soviet Union bankruptcy formally ended the Cold War. As Boris Gubanov wistfully told James Hartford:

Energia's future is bleak, if not used for the [ISS] *Space Station.*

As it turned out, Russia came out pretty good on ISS for a long period after the U.S. Space Shuttle decommissioned and NASA was forced to purchase tickets from them to deliver our astronauts there and back aboard the Soyuz vehicle until commercial crew and cargo transportation services finally became available.

Chapter Fourteen: Hard Lessons; Tough Decisions

HUMANKIND'S PREVIOUS BRIEF experiment with space exploration both reflects and predicts conflicting prospects for peril and promise. Our wisdom and willingness to confront great challenges will continue to shape national and global progress; will influence individual qualities of life and opportunity; and will determine the substantive value of lessons and legacies we bequeath to generations who follow.

Although the Mercury-Gemini-Apollo programs reaped indisputably great American prestige, Kennedy's predecessor President Eisenhower had believed that achievements undertaken primarily for that purpose were ill-advised. Many military leaders agreed, arguing that Project Apollo diverted funding away from more important national defense priorities. The decision to race the Soviet Union to the Moon even reversed Kennedy's original preference upon entering the White House. In his inaugural address he had suggested "let us explore the stars together," and until just days before his assassination, he was still pursuing a possible cooperative lunar landing prospect with the Soviet Union.

Was Kennedy really serious about such a proposal? If not, it seems unlikely that he would have presented a proposition before a September 20, 1963 UN General Assembly asking:

Should Man's first flight to the Moon be a matter of national competition?

…then suggesting that the U.S. and Soviet Union explore the possibility of:

…a joint expedition to the Moon.

By 1963, the U.S. had established strategic leverage over the USSR through demonstrations of superior economic, technological and military power. This occurred as the Soviet Union first backed off a confrontation in 1961 over access to Berlin, and again in yielding to U.S. pressure to remove its missiles from Cuba in October 1962. Kennedy proposed a strategy of peace to reduce serious nuclear war tensions which led to superpower signing of a Limited Test Ban Treaty in August 1963.

Human space programs played an important symbolic role in lessening those tensions as global audiences watched TV images on July 17, 1975 showing Soviet cosmonauts and NASA astronauts embracing high above the Atlantic Ocean during the historic docking together of the American Apollo Command and Service Module and the Russian Soyuz vehicle.

Yet in the absence of a Kennedy-like commitment or clearly defined national goals, the U.S. space exploration program is broadly viewed by the public as a discretionary activity. As a result, NASA has become, as the 2003 Columbia Accident Investigation concluded, "an organization straining to do too much with too little."

Our nation has an important leadership decision to make. Just as Kennedy asked more than a half century ago, are we prepared to accept leadership burdens which "will last for many years and carry very heavy costs," bluntly pointing out that "it would be better not to go at all" if we are not "prepared to do the work and bear the burdens to make it successful."

In considering our response, perhaps it is valuable to contemplate how future generations might respond to the archetypical Apollo legacy question "If we can put a Man on the Moon, why can't we: (fill in the blank)?" Let's not give our descendants reasons to ask of us, "Why did you stop there?"

Any effective policy requires committed leadership, a sustainable strategy, and a carefully conceived long-term roadmap with achievable, worthwhile milestones. Since none of these prerequisites presently exist, the U.S. Space program is adrift without any real vision, goal, or determination.

Over its more than six-decade history, NASA has become a mature bureaucracy, complete with oppressive management redundancy, political baggage, and organizational inertia which would have hamstrung the success of groundbreaking achievements during the agency's glory days.

The agency's continuing economic and program viability relies upon a coalition of local and regional politicians together with aerospace industry interests and an institutional and facility base stretching between Florida, Louisiana, Texas, Alabama, Utah and California. As the Challenger accident investigation revealed, resulting rivalries and misunderstandings can contribute to terrible consequences.

Bob Crippen, a four-time Shuttle astronaut, Kennedy Space Center operations deputy director from 1986 to 1989 and Challenger investigation participant reported:

> One thing that came out of the Challenger investigation was that between Marshall [Space Flight Center] and Johnson and NASA headquarters the communication was poor.

He observed that:

> Some people knew stuff at one place that people at another didn't know. That night when it was so cold, people at the contractor were saying it was too cold for solid rockets to fly. Marshall Space Flight Center knew that Johnson Space Center did not know that. One of the things we worked hard to improve was communication.

Allan McDonald, Morton Thiokol's top official at Cape Canaveral for the Challenger launch attributes at least some of that communications failure to NASA center rivalries. As also discussed in his book *Truth, Lies and O-Rings: Inside the Challenger Disaster*, he observes:

> Immediately after the Challenger accident, [he] heard that people at Johnson directed a lot of anger at Marshall. Marshall and Johnson were competing with each other for a share of the shuttle program and a share of the work. That led to people failing to share information. The Marshall folks would have told the mission management team about their discussion with Thiokol.[77]

McDonald concludes:

> That might have made the mission management team cancel the launch.

Part Four: Lost in Space

NO RECENT PRESIDENT has established a future vision for space exploration with any real promise to achieve long-term political and budgetary support vital for success.

As the Space Station era which replaced Apollo now approaches its own end, proposals to return to the Moon, harvest lunar and asteroid resources, and/or conduct expeditions to Mars and beyond lack unified priorities and commitments.

A 2014 National Research Council report prudently observes that such ambitious plans cannot succeed…

> …*without a sustained commitment on the part of those who govern the nation—a commitment that does not change direction with succeeding electoral cycles. Those branches of government—executive and legislative—responsible for NASA's funding and guidance are therefore critical enablers of the nation's investment and achievements in human spaceflight.*[78]

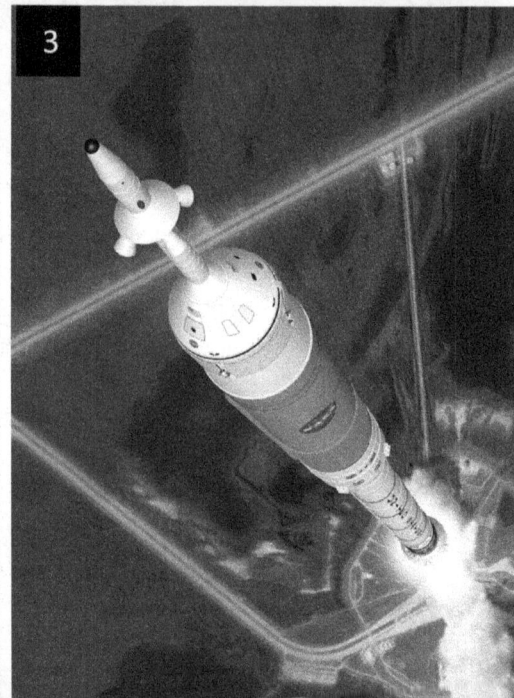

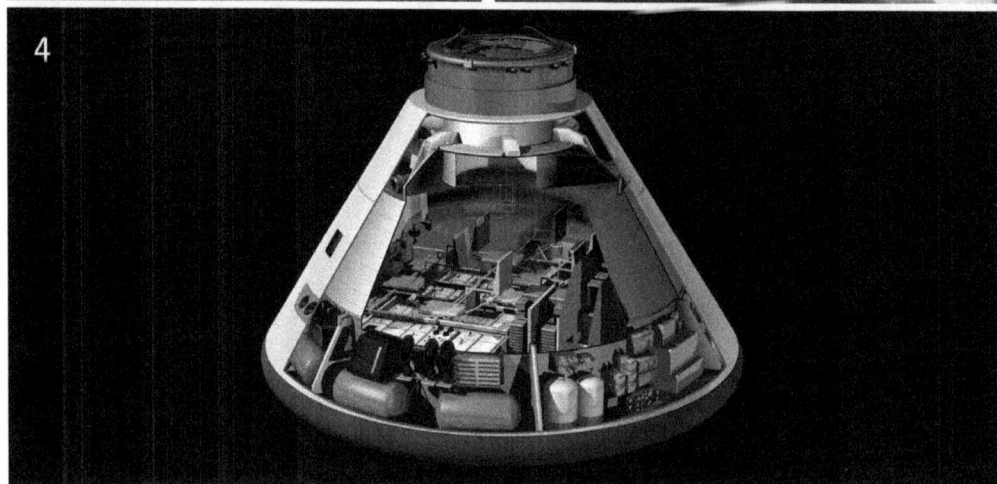

1. Barack Obama
   Source: NASA

2. Ares V Rocket
   Source: NASA / MSFC

3. Ares I Rocket
   Source: NASA

4. Orion Crew Exploration Vehicle
   Source: NASA

1. Altair Lunar Lander
   Source: NASA
2. Orion MPCV
   Source: NASA

3. Aerojet Rocketdyne RL 10
   cryogenic engine used in many U.S.
   rockets. Source: NASA

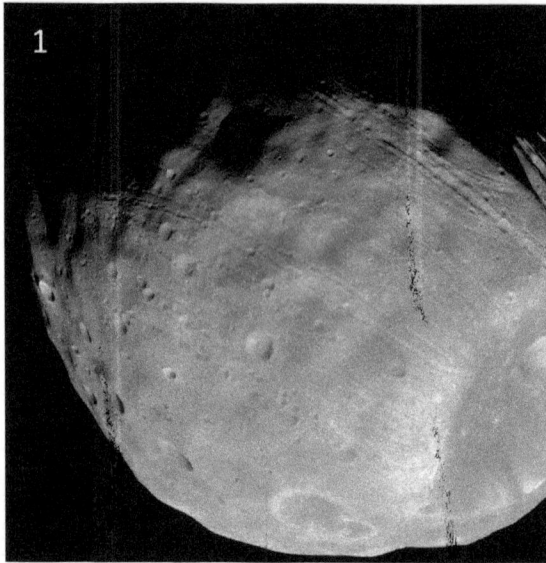

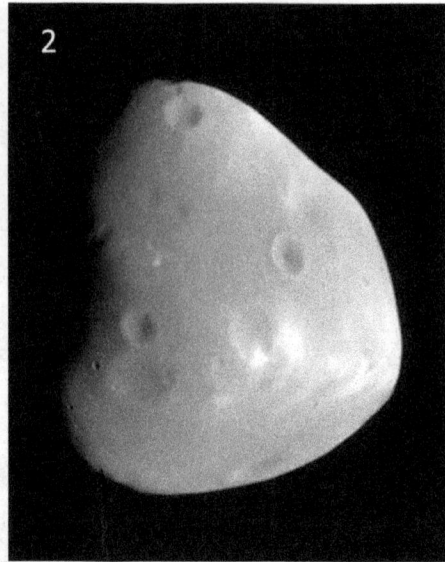

1. Phobos
   Source: NASA
2. Deimos
   Source: NASA
3. Space Launch System
   Source: NASA
4. Bonnie Dunbar
   Source: NASA
5. George W. Bush
   Source: NASA

Chapter Fifteen: Life and Death of the Constellation Program

THE GEORGE W. BUSH administration envisioned a Constellation Program (2005-2009) which would borrow and extend concepts from Apollo and Space Shuttle programs for broader lunar and deep space applications. Major purposes were to complete assembly of the ISS, return to the Moon no later than 2020, and ultimately, to achieve crewed flight to Mars beginning in the 2030s as an ultimate goal. Technical aims included expanding astronaut experience beyond low-Earth orbit and developing new technologies to enable sustained human presence on other planetary bodies.

Constellation goals laid out in Vision for Space Exploration under NASA Administrator Sean O'Keefe and formalized by NASA Authorization Act of 2005 directed NASA to:

> …develop a sustained human presence on the Moon, including a robust precursor
> program to promote exploration, science, commerce and US preeminence in
> space, and as a steppingstone to future exploration of Mars and other destinations.

The private non-profit National Space Society (NSS) strongly supported a permanent return to the Moon for the purpose of creating new industries. The Bush administration agreed, arguing that extended human presence on the Moon would reduce costs of future space exploration by "harvesting and processing lunar soil into rocket fuel or breathable air," "developing and testing new approaches and technologies and systems," and thereby establishing a sustainable course of long-term exploration.

NASA's website[79] highlighted reasons for going to the Moon to include:

- To extend colonization.

- To further pursue scientific activities intrinsic to the Moon.

- To test new technologies, systems, flight operations and techniques to serve future space exploration missions.

- To provide a challenging, shared and peaceful activity to unite nations in pursuit of common objectives.

- To expand the economic sphere while conducting research activities that benefit our home planet.

- To engage the public and students to help develop the high-technology workforce that will be required to address the challenges of tomorrow.

Although the lunar plan—along with other key elements—were subsequently cancelled by the Barack Obama administration, Key Constellation program elements would have included a new Ares I launcher and larger Ares V cargo booster to crews and cargo (respectively) to the ISS, an Orion Crew Exploration Vehicle (OCEV), and an Altair lunar lander.

The OCEV to be developed by Lockheed Martin would have consisted of three main parts: A Crew Module (CM) similar to Apollo Command Module but capable of supporting a crew of six (later cut back again to four to save money); a cylindrical Service Module (SM) containing primary propulsion and supplies; and a Launch Abort System (LAS) for emergency escape during ascent. Crew Module would be designed for 10 flights.

An Orion variant now called the Orion Multipurpose Crew Vehicle (MPCV) was salvaged from the Constellation program and continues.

The Altair lander was to have two parts: an ascent stage to house a four-person crew; and a descent stage with storage for the majority of crew consumables (oxygen and water) plus scientific equipment. Powered by four RL-10 rockets used in Centaur upper stage of Atlas IV, Altair was designed to land in a lunar polar region where surface water might be accessed.

Like Apollo, Altair would use lunar orbit rendezvous, but unlike Apollo, the crew would be launched separately for Earth orbit rendezvous aboard Altair. Also, unlike Apollo, the Orion vehicle would remain unmanned in lunar orbit.

NASA estimated Constellation cost, including a separate ISS Commercial Crew and Cargo program, to be $230 billion (2004 dollars) through 2025. Upon taking office, the Obama administration determined that the program was over budget and behind schedule and that neither a return to the Moon or manned flight to Mars was affordable. Accordingly, Constellation program funding was eliminated from NASA's 2011 Federal budget.

There would be no budget for lunar landers or bases.

Chapter Sixteen: A "Flexible Path to Mars"

AT AN APRIL 15, 2010 Space Conference in Florida, President Obama and top administration officials unveiled a Flexible Path to Mars option based upon a recommendation of a study committee panel headed by former Lockheed Martin CEO Norman Augustine.

Following plans to retire the Space Shuttle, the panel had been charged with responsibility to determine:

> ...a vigorous and sustainable path to achieving its boldest aspirations in space.

Key objectives assigned for committee review included ongoing support for the International Space Station, development of missions beyond low-Earth orbit (including the Moon, Mars and near-Earth Objects), and use of the commercial space industry. More specific parameters for consideration included:

> ...crew and mission safety, life-cycle costs, development time, national space industrial base impacts, potential to spur innovation and encourage competition, and the implications and impacts of transitioning from current human space flight systems.[80]

The Augustine Committee concluded that the ultimate national goal for human space flight should require both physical and economic sustainability. Also important are to support international cooperation, develop new industries, advance energy independence, promote national prestige, and reduce climate change.

Potentials for obtaining resources such as water (a source of oxygen for breathing and hydrogen to combine with oxygen for rocket fuel), and other resources of possible construction or even industrial value were associated with most ideal destinations. Although there was no specific reference to harvesting in-situ resources from asteroids, the report did mention the possibility of evaluating near-Earth objects for *their utility as sites for mining of in-situ resources.*

While noting difficulties of travel into deep gravity wells of the lunar and Martian surface in order to access precious resources, the Committee concluded that:

> Mars stands prominently above all other opportunities for exploration.

Further, the report observed:

> Mars is unquestionably the most scientifically interesting destination in the inner

Solar System, with a history much like Earth's. It possesses resources which can be used for life support and propellants. If humans are ever to live for long periods on another planetary surface, it is likely to be on Mars.

The Committee qualified its advocacy for Mars by pointing out that the Red Planet is:

...not an easy place to visit with existing technology and without a substantial investment of resources.

Accordingly, it advised that:

Mars is the ultimate destination for human exploration, but it is not the best first destination.

The Augustine Committee's final report outlined three basic options for Obama administration's consideration, clearly favoring the third as most promising. The first would begin with a brief test of equipment and procedures on the Moon followed by a Mars landing. The second would involve more extended lunar surface operations focused upon developing capabilities to explore Mars.

The transparently preferred option would provide a Flexible Path to inner Solar System locations. The first destinations would include lunar orbit, Lagrange points (locations in an orbital plane between two bodies where a spacecraft will remain in gravitational equilibrium), near-Earth objects, and Mars' moons Phobos and Deimos. Those moons could serve as locations to coordinate or control robots on the Martian surface. These precursor missions would be followed by exploration of the lunar and/or Martian surface(s), potentially involving development of propellant depots.[81]

The Augustine Committee strongly favored having crew transport services for ISS passed from NASA to the private sector. Speaking at a Washington, DC press conference, Chairman Augustine said:

We think this is a time to create a market for commercial firms to transport both cargo and humans between the Earth and low-Earth orbit. While that is certainly not simple, it is much easier than going to Mars. We think NASA would be better served to spend its money and its ability—which is immense—focusing on going beyond low-Earth orbit rather than running a trucking service to low-Earth orbit.

Given then-current time schedules, committee members were critical of a plan to produce the Ares I and Ares V rockets along with the Orion crew ship.

They regarded it a paradox for Ares I and Orion to be planned to enter service as Shuttle replacements in 2017...one year after the ISS program was scheduled for termination. In addition, since no lunar landing capability would be ready by then, Orion would have nowhere to go.

In response to Augustine Committee suggestions, the Obama administration cancelled Ares I and V and the Altair lunar lander altogether, redirected Orion development to become a Multi-Purpose Crew Vehicle (Orion MPCV) for destinations beyond low-Earth orbit, and approved development of a new heavy-lift Space Launch System (SLS) to support human and cargo Flexible Path missions into the inner Solar System.

The initial SLS version will be powered by four of 16 left-over Space Shuttle main engines and upgraded Shuttle-heritage solid fuel boosters which, altogether, are about 10 percent more powerful than the Apollo Saturn V. A later version using more advanced strap-on boosters and a higher-energy upper stage is aimed at advancing from a 70-metric-ton launch capability to a 130-metric-ton capacity.

It's notable to realize White House approval of the SLS came about only after intense pressure from congressional advocates in key impacted districts, while at the same time, no missions to Mars or elsewhere are currently funded by Congress or even in detailed NASA planning stages.

Nor have technical requirements for such missions yet been specified. Instead, the decision to proceed with development of a new heavy-lift rocket came from an internal NASA review known as Key Decision Point C, in which the agency says:

> ...*provides a development cost baseline for the 70-metric-ton version of the SLS of $7.021 billion from February 2014 through the first launch and a launch readiness schedule based on an initial SLS flight no later than November 2018.*[82]

Meanwhile, the Obama administration proposed to feature an SLS mission with no scientific or technical planning foundation whatsoever.

Chapter Seventeen: Chasing Asteroids

AS PREVIOUSLY DISCUSSED, although the Augustine Committee's flexible pathway report made no specific reference to harvesting precious resources from asteroids, it did suggest the possibility of evaluating near-Earth objects for their utility as sites for mining of in-situ resources.

Apart from presently unknowable technical and economic arguments for this, a major question revolves around plausible timing.

Asteroids are much further away than the Moon—which influences access and control communications; are in far less predictable orbits and spin conditions; are far less well understood with regard to resource composition and surface characteristics affecting mining yields and operations; are much smaller trajectory targets; and lack gravity so must be docked-with and anchored-to rather than landed on.

Materials obtained must be of inordinate mega-value to offset astronomically high access, collection, processing, return transit costs and business risks in order to compete with far more predictable and less expensive Earth-sourced alternatives. It is most unlikely that even the precious metals at today's market prices would afford promising profitability prospects even at dramatically lower space transportation costs.

For reference, a NASA Stardust mission which brought back microscopic comet dust particles, reportedly cost $200 million. The only other, a Japan Aerospace Exploration Agency probe, collected microscopic asteroid particles. Notably, neither of these missions demanded complex collection or processing procedures. Determining a cost-effective way to bring back exponentially larger commercial quantities or acquire practical amounts for in-space use will present enormous technical and investment challenges.

As for accessing various resource options, C-type asteroids are believed to contain an abundance of water, and carbon; S-type may contain numerous metals including gold, platinum and rhodium; and M-types are far rarer, but likely contain many times more metals than S-types. Harvesting operations may involve surface mining using scoops, augers or magnetic rakes for certain metals; heating to vaporize and collect volatile materials such as water; and shaft digging which requires accurate astro-location and means to extract, separate and transport materials.

Space processing will be very tricky. Very low gravity conditions will require machinery to somehow be anchored in place—which may be particularly difficult if the surfaces are primarily formed of loose rubble; operations will need to be highly automated and maintenance free under extreme thermal conditions.

Once obtained, the materials must then be separated, processed and made accessible for use. One possible approach would be to bring raw asteroid materials all the way to Earth. Another would be to process desired material on-site and then bring it back and/or use some for propellant. At least as a scientific starter, the Obama administration had proposed a third.

After cancelling lunar plans of the previous White House, President Obama tasked NASA to prepare to deploy an unmanned robotic ship powered by solar-electric propulsion to capture an asteroid and drag it into lunar orbit for four astronauts transported by the SLS on an Orion MPCV to inspect. Termed the Asteroid Redirect Mission, this was proposed to be accomplished by 2021.

According to Donald Yeomans, who heads NASA's Near Earth Object (NEO) program, the idea was to select a suitable 25-foot, near-Earth candidate asteroid (from less than 10 million miles away) which would be captured with the space equivalent of a baggie with a draw string.

After it is bagged…

> *You attach the solar propulsion module to de-spin it and bring it back where you want it.*

Yeomans argued that a 25-foot asteroid drifting off course poses no threat because it would burn up if it inadvertently entered Earth's atmosphere.

There is small likelihood that a Republican-dominated Congress or White House would support funding for such a venture which the Keck Institute for Space Studies has estimated through a similar proposal would cost about $2.6 billion. George Washington University Space Policy Institute Director Scott Pace, a high-level NASA official during the George W. Bush administration, believes it is a bad idea both from scientific and international cooperation standpoints.

Five-mission NASA astronaut Bonnie Dunbar notes that while planetary protection from asteroid impacts has been part of NASA and DOD research and robotic mission portfolios for many years, the Obama asteroid redirect plan was poorly conceived and had never been coordinated with NASA or peer reviewed either from scientific or engineering perspectives prior to being announced in a public press release.

She points out:

> *Most importantly, while we know where the Moon and Mars are and can predict their orbits for planning purposes, this is certainly not the case with a yet-to-be identified target asteroid. However, this lack of information did not deter the Obama administration from identifying a date for this mythical rendezvous: as early as 2021.*

Dunbar points out our traditional international governmental space partners, who are critical for spreading the funding risk involved with future exploration, hadn't embraced the plan.

She stated:

> *This was very evident at a recent Association of Space Explorers meeting in Cologne, Germany, when ESA leadership stated that these missions could be more effectively performed by robotic spacecraft, and that they did not plan to include them in their strategies.*[83]

Yet, if the asteroid redirect plan was a poorly conceived science agenda, it gets much worse.

Part Five: Launching Private Enterprise

IN THE EARLY 1980s, the idea that a private company would infringe upon the exclusive domain of governments to launch rockets into space seemed quite audacious to most people. Nevertheless, Houston-based Space Services Inc. of America (SSIA) founded by the late David Hannah, a real estate developer, and headed by former Mercury astronaut Deke Slayton, not only believed it could be done, but actually accomplished it. On September 9, 1982, SSIA's Conestoga I rocket delivered a 1,000-pound (450 kg) dummy payload (mostly water) to an altitude of 194 miles (313 km)…more than three times higher than needed to actually qualify as reaching space.[84]

SSIA began their enterprise with a bold notion that private enterprise could lower the cost of space launches by clustering a variable number of small engines around a relatively inexpensive expendable booster as necessary to accommodate specific launch payload requirements. The 1982 launch used surplus engines SSIA acquired from the second stage of Minuteman missiles. A later design used surplus 1960s-vintage Scout missile engines.

EER Systems purchased SSIA in 1990 with plans to market launch services to NASA for their Commercial Experiment Transporter (COMET) microgravity research program. Westinghouse signed on to provide control systems, Space Industries, Inc. was to develop the re-entry module, and EER was contracted to supply several Conestoga launchers. A series of COMET program delays, budget overruns, and an October 23, 1995 Conestoga failure 46 seconds into the flight ended EER's business with NASA. The company's remaining assets were purchased by L-3 Communications in 2001 for $110 million.

In November 2005, NASA initiated a new Commercial Orbital Transportation Services (COTS) program involving a number of non-binding Space Act Agreements providing milestone-based payments aimed at attracting private sector investment and innovation to substantially lower the price of access to space. Real progress is evidenced by a growing variety of existing and entirely new companies that are providing International Space Station cargo resupply services, developing round-trip crew transport and emergency return capabilities, and bending technology advancement trajectories to higher orbits.

As articulated by NASA Administrator Michael Griffin:

> With the advent of the ISS, there will exist for the first time a strong, identifiable market for "routine" transportation service to and from LEO, and that this will be only the first step in what will be a huge opportunity for truly commercial space enterprise. We believe that when we engage the engine of competition, these services will be provided in a more cost-effective fashion than when the government has to do it.

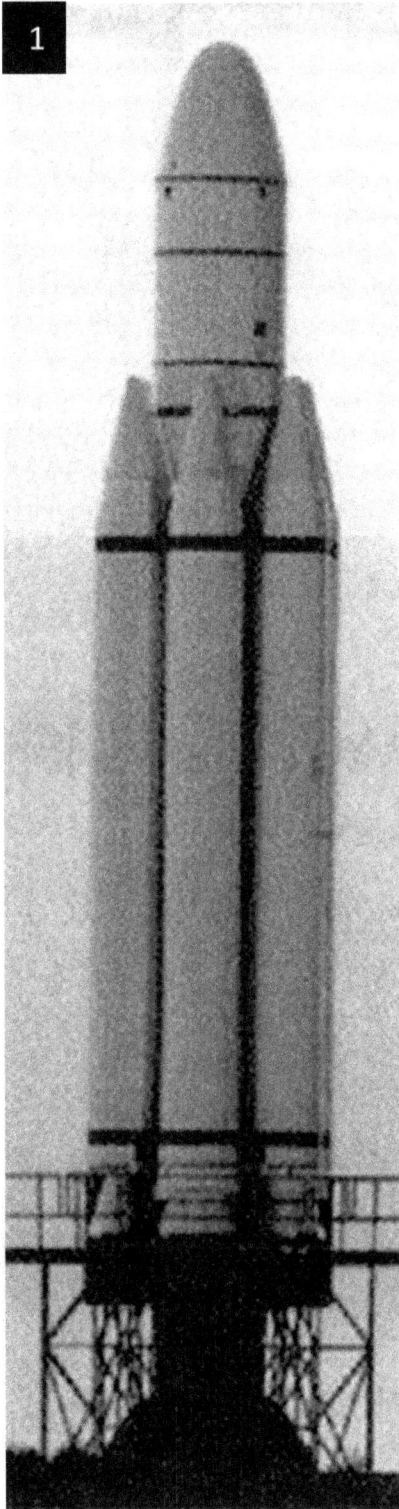

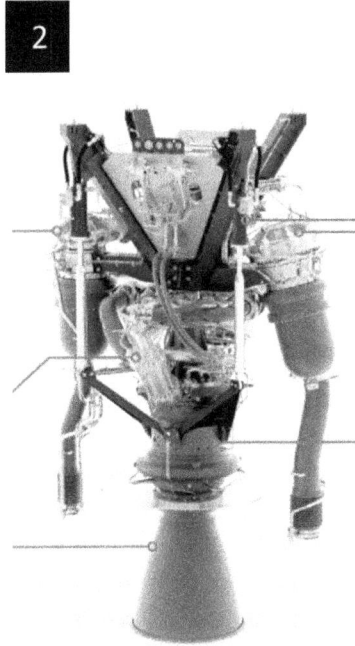

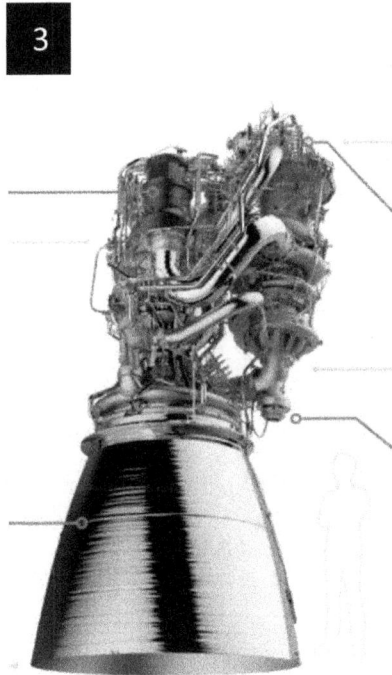

1. Conestoga 1 Rocket
   Source: NASA
2. BE-3 Engine
   Source: Blue Origin

3. BE-4 Engine
   Source: Blue Origin
4. New Glenn Rocket
   Source: Blue Origin

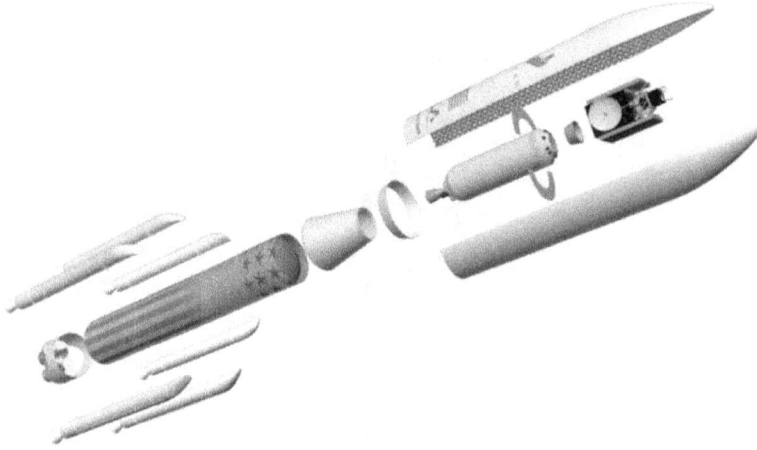

1. Vulcan Rocket
   Source: NASA
2. Falcon 1 Rocket
   Source: SpaceX
3. Starship Prototype
   Source: SpaceX

4. Merlin 1C Engine
   Source: SpaceX
5. Boeing Starliner
   Source: NASA

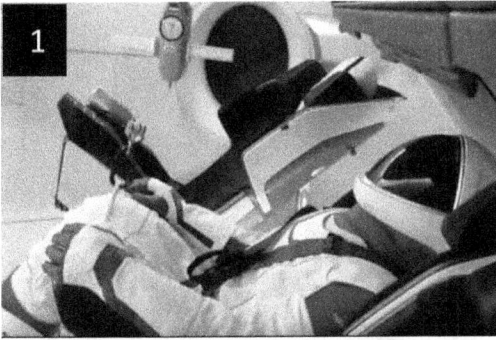

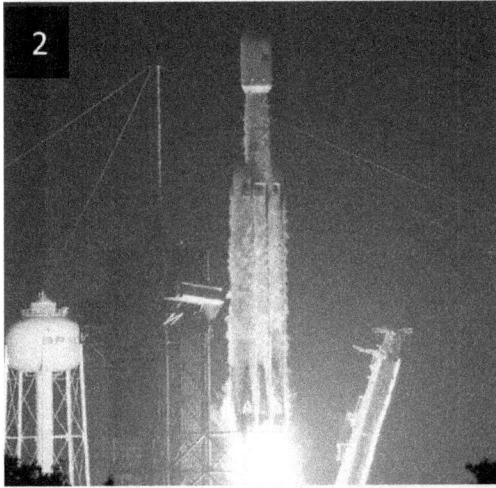

1. Ripley the mannequin
   Source: SpaceX
2. Falcon Heavy Rocket
   Source: NASA
3. Ariane 5 Rocket
   Source: DLR
4. Demo-2 Mission astronauts
   Source: NASA

1. Soviet Heavy lift Proton Rocket
   Source: NASA
2. Antares Rocket
   Source: NASA
3. SpaceX Raptor Engine
   Source: Brandon De Young
4. Starbase
   Source: Drive Tesla Canada

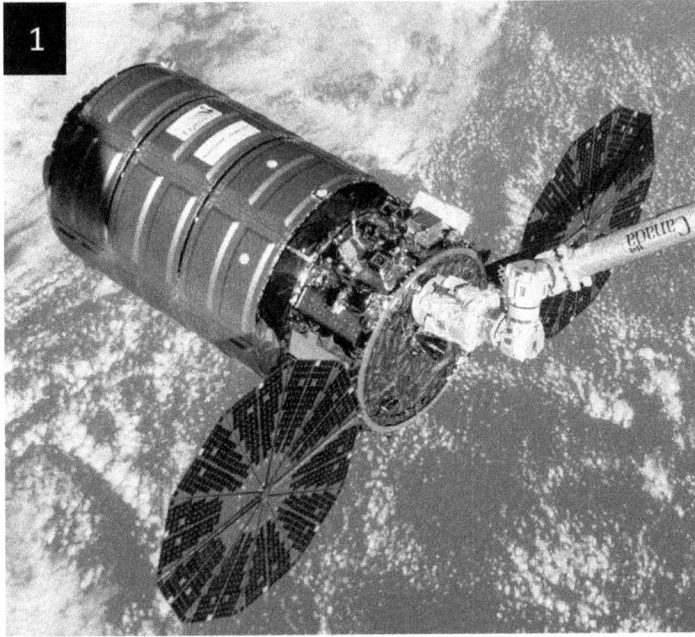

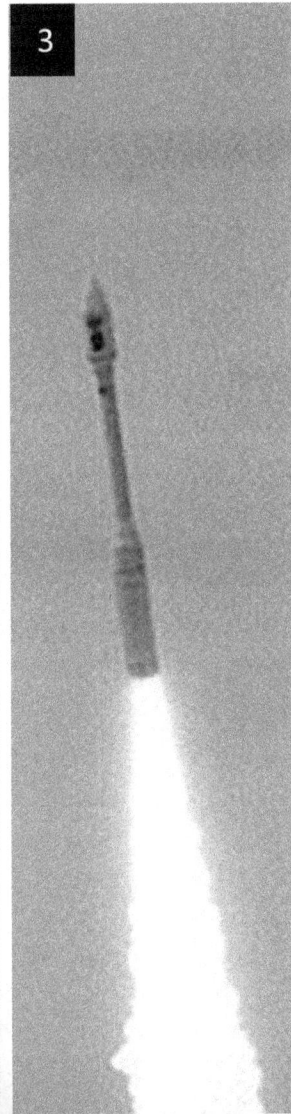

1. Cygnus Spacecraft
   Source: NASA

2. Russian RD-181 rocket engines used in the Antares rocket.

3. Minotaur C
   Source: Public Domain

4. Lunar Gateway Concept
   Source: NASA

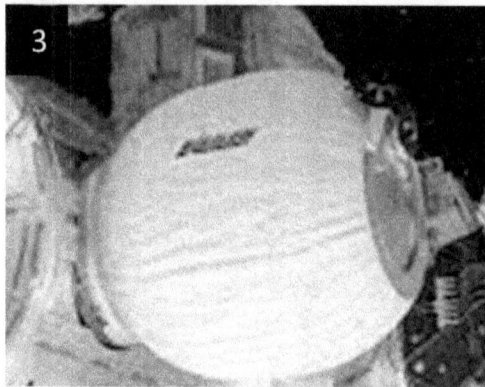

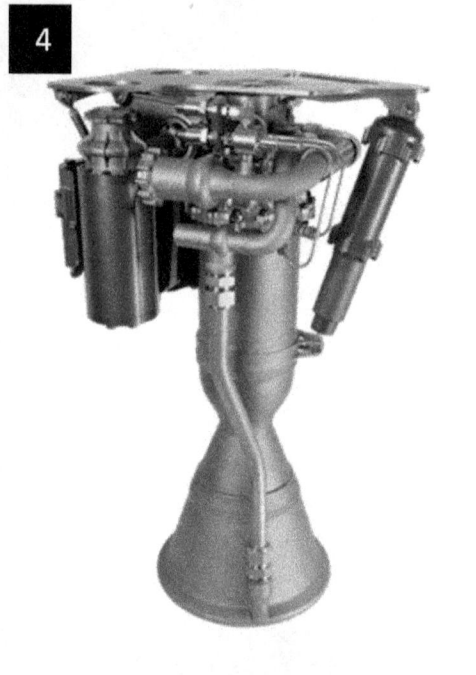

1. Sierra Nevada Dream Chaser lifting body space plane. Source: NASA
2. Boeing X-20 Dyna Soar Source: Public Domain
3. Bigelow Expandable Activity Module (BEAM). Source: NASA
4. Rocket Lab's Rutherford Engine used in the Electron rocket. Source: RocketLab
5. Electron Rocket Source: Public Domain

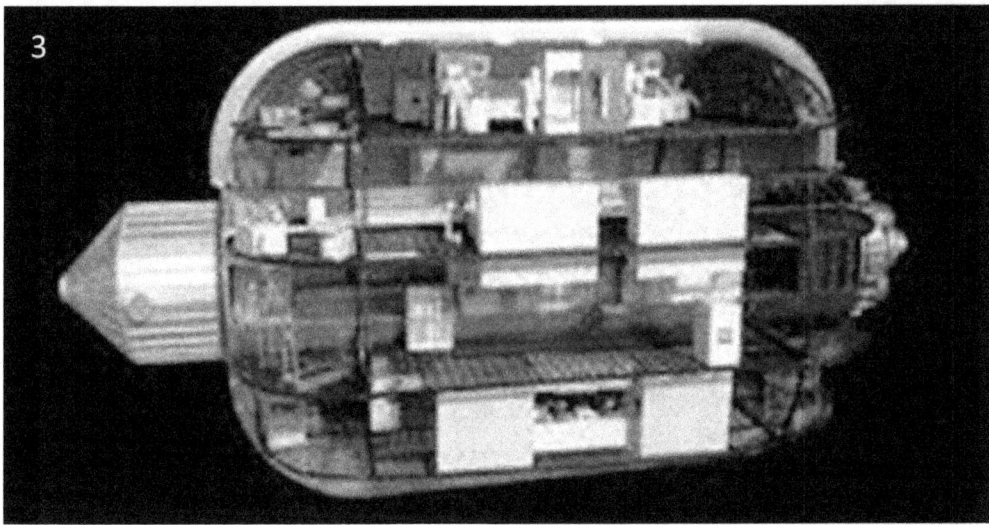

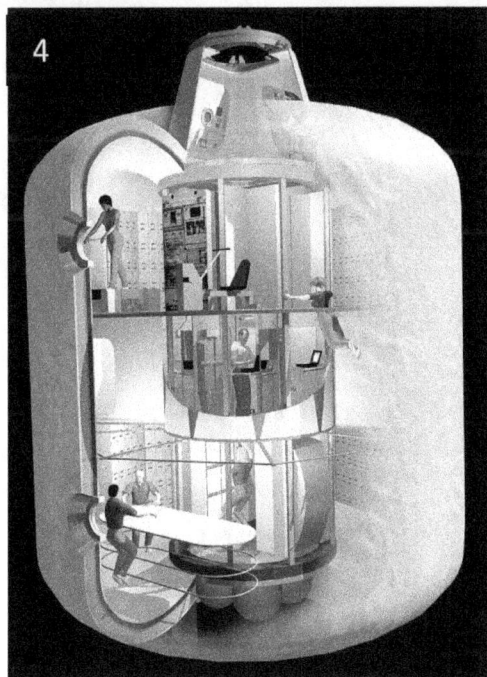

1. B-330
   Source: Bigelow Aerospace
2. Robert Bigelow
   Source: NASA
3. B-2100 "Olympus"
   Source: Bigelow Aerospace
4. Transhab
   Source: NASA

1. Space Ship Two
   Source: Virgin Galactic
2. Burt Rutan, American aerospace
   engineer and entrepreneur.
   Source: D Ramey Logan
3. White Knight 2
   Source: Virgin Galactic
4. SpaceShip One
   Source: Virgin Galactic

Chapter Eighteen: The New Entrepreneurs

THE U.S. CONGRESS and Pentagon have been anxious to end heavy reliance upon Russian RD-180 rocket engines, particularly for national security-sensitive launches. A key strategy in accomplishing this is to award competitive contracts that incentivize new developments by American engine and rocket companies.[85]

Special emphasis on this priority can be traced to a 1995 U.S. Air Force Evolved Expendable Launch Vehicle (EELV) program which anticipated supportive commercial launch market investments to offset government costs. At that time the Air Force contributed $1 billion to encourage McDonnell Douglas (now Boeing) to develop a heavy-lift Delta IV rocket derived from the former Delta III, and for Lockheed Martin to refine their Atlas V to provide a more powerful first stage.

The plan was to allow Lockheed Martin to continue to import RD-180s under the condition that they would eventually be manufactured in the U.S. through a coproduction arrangement. A fiscal 2017 Authorization Act capped new RD-180 imports at 18 total through 2022.

The Pentagon's EELV program was initiated in response to an important 1994 congressionally directed study that concluded that the RD-180 was attractive for application with the Atlas V due to comparative off-the-shelf availability and its use of conventional RP-1 rocket-grade kerosene fuel. Delta IV vehicles were determined to be more expensive but providing two rocket options was considered to be important for national security reasons.

As it turned out, the commercial EELV launch market never materialized as hoped. Nevertheless, a 2006 joint venture between Boeing and Lockheed Martin which created the United Launch Alliance (ULA) offered the Air Force two rocket options from a single provider.

Russian annexation of Crimea in 2014 aroused high levels of U.S. government concern about U.S. dependence upon RD-180 use. In 2015, Congress directed the Air Force to again create private rocket engine technology which would wean away their use through competitive contracts. In response, the Air Force awarded $242 million in contracts to Aerojet Rocketdyne, Orbital ATK, SpaceX and ULA stipulating that at least one-third of total cost of each prototype project be privately funded.

Blue Origin's BE-3 engine, which powers their New Shepard sub-orbital rocket, is the first new liquid hydrogen-fueled rocket engine to be developed for production in America in over a decade.

The BE-4 engine, currently under development by Blue Origin, is the first oxygen-rich staged combustion engine made in the U.S. It will power the next generation of American orbital rockets like the New Glenn as well as ULA's Vulcan launch vehicles.[86]

SpaceX has successfully developed its Merlin series engines—which power the Falcon launch vehicles—and are currently developing the Raptor engine which will power the Starship into LEO and beyond.[87]

Space Race to the Space Station and Beyond

U.S.-produced rocket engines had previously represented only incremental modifications to those designed in the 1960s with the exception of Space Shuttle Main Engines (SSMEs). NASA's Space Policy in 1982 declared the Shuttle as the primary space launch system for both the United States national security and civil missions.

Plans to decommission the vehicle by 2011 changed all that.

Beginning in 2008, NASA has competitively awarded contracts for private ISS cargo services through its Commercial Resupply Services (CRS) program, along with funding to develop capabilities to achieve safe and reliable crew rotation and emergency return services through separate Commercial Crew Development (CCDev), Commercial Crew Integrated Capability (CCiCap), and Commercial Crew Transportation Capability (CCtCap) contracts. These contracts engage risk-tolerant billionaire space entrepreneurs and seasoned, publicly traded aerospace entities.[88]

Elon Musk's SpaceX Developments

Founded in 2002 by PayPal co-founder Elon Musk, Space X (the "X" referring to Exploration Technology), is a big competitor for two-way ISS cargo and crew transfer business applying a reusable rocket design. Its signature, two-stage, 180 feet tall Falcon 9 booster carries a Dragon capsule capable of carrying up to seven passengers to the ISS where it is grappled by a remote manipulator system and attached to a berthing port.

Launch concept tests beginning in 2012 soft-landed Musk's self-funded Grasshopper Vertical Takeoff, Vertical Landing (VTVL) rocket prototype multiple times following brief flights that did not reach space. SpaceX also soft-landed the first stage of its Falcon 9 Reusable rocket on the ground, although not in a fully operational mode. The booster was flown to an altitude of only 820 feet with its landing legs already deployed at launch and no payload on top.[89][90][91]

The company has already logged several historic achievements: in 2008, SpaceX became the first privately funded company to launch a (Falcon 1) rocket into orbit; the first to successfully orbit and recover a spacecraft (2010); the first to send a spacecraft to the ISS (2012); the first to launch a satellite into LEO (2013); and the first organization, private or government, to successfully return a first stage back to the launch site and accomplish a vertical landing with a rocket on an orbital trajectory (2015). SpaceX also conducted its first satellite delivery to geosynchronous orbit (NASA's Deep Space Climate Observatory) in 2015.[92][93]

After launching an unmanned Dragon capsule to the ISS, SpaceX successfully demonstrated reuse of Falcon's first rocket stage with a perfect April 9, 2016 landing on a floating barge. Following four previously missed attempts, this major achievement has since been repeated along with launches with reused rockets.

That history includes some disappointing setbacks—including a recent one. Elon Musk described a catastrophic September 1, 2016 Falcon 9 rocket explosion during a routine refueling exercise as:

> ...turning out to be the most difficult and complex failure we have ever had in 14 years.

Although no one was injured, it raised safety procedure concerns.

While the SpaceX refueling process—which involves cooling liquid oxygen to just 22 degrees or so above the freezing point (-362 degrees Fahrenheit) to improve engine thrust, the short preparation time before launch requires that astronauts be in the vehicle before that fueling occurs. This requirement conflicts with traditional methods where the astronauts board after it is clear that fueling has occurred safely. In December, retired Air Force Lt. General Thomas Stafford stated that the SpaceX procedure is:

> ...contrary to booster safety criteria that has been in place for over 50 years, both in this country and internationally.

In September 2016, a Falcon-9 launch vehicle and the payload were lost in a launch pad explosion during propellant filling procedures prior to a static fire test. The event also came a year before SpaceX hoped to launch a first test flight to the ISS with astronauts as part of NASA's Commercial Crew Development (CCDev) program. This was planned in competition with Boeing's CST-100 Starliner capsule carried on United Launch Alliance's Atlas V rocket which has not suffered a major failure in more than 60 missions.

At the time of this writing, since this September 2016 accident, SpaceX has launched numerous consecutive successful missions to LEO and beyond on its Falcon rockets. A vast majority of the first stages were successfully landed on drone ships in the ocean, brought back to land and reused for subsequent launches.

In May 2021, the Falcon 9 Booster B1051 launched and landed for the tenth time, achieving one of SpaceX's milestone goals for reuse. Additionally, SpaceX has also successfully recovered fairings and reused them multiple times.[94]

In 2013, a SpaceX Dragon spacecraft destined for ISS arrived and docked a day late following thruster control problems which were corrected from the ground. Much worse, on June 28, 2015, a Falcon 9 topped with an unmanned Dragon capsule which was to deliver supplies to ISS exploded as the result of a broken strut which caused an upper-stage oxidizer tank on the 15-story-tall booster to rupture.

In May 2017, SpaceX launched a top-secret National Reconnaissance Office (NRO) satellite, its fifth launch of that year. One week later the company accomplished what it described as a major new milestone in completing its first static fire of the center core of "…the most powerful booster rocket since NASA's Saturn V Moon rocket," a Falcon 9 Heavy with 27 engines. Then, only one week after that, a Falcon 9 Heavy successfully delivered a Boeing-made 6.7-ton Inmarsat Global Xpress satellite to orbit.

SpaceX announced opportunities to average about two launches per month throughout 2017, possibly increasing to weekly launches by 2019. SpaceX has also been actively building a satellite internet constellation in LEO with the Falcon 9 vehicles, launching up to 60 satellites at a time. They aim to deploy 1,584 satellites that will provide near global Internet service by late 2021 or 2022.[95]

Headquartered in the Los Angeles suburb of Hawthorne, California, SpaceX currently has an estimated employment of about ten thousand people. Its large, three-story facility, which was originally built by Northrop Corporation to build Boeing 747 fuselages, houses the company's offices, mission control and vehicle factory. SpaceX also has regional offices in Houston, Texas; Chantilly, Virginia; Redmond, WA, and Washington, DC.[96]

Company growth has been rapid. NASA first awarded SpaceX a $278 million COTS seed money contract in 2006 to demonstrate a cargo delivery capability for the ISS with a possible option for crew transport. NASA had originally also signed a COTS agreement with Rocketplane Kistler (RpK) in 2006, but later terminated the agreement due to RpK's insufficient private funding. SpaceX launched and orbited its first successful COTS demonstration flight on December 9, 2010 after a two-year delay. A second COTS demonstration Dragon capsule flight made a rendezvous with ISS but intentionally didn't dock.

SpaceX was awarded a $1.6 billion Commercial Resupply Services contract in 2008 to provide 12 unmanned cargo deliveries to the ISS through 2016, while the Orbital Science Corporation received $1.9 billion to provide nine missions. The 2008 award was later extended to provide 20 flights for SpaceX and 10 for Orbital Science.[97][98]

In 2011, NASA awarded SpaceX an additional $75 million to develop their launch escape system and further advance progress on Falcon 9/Dragon design under its Commercial Crew Development program. In 2012 NASA awarded SpaceX (along with Boeing and Sierra Nevada) a contract of up to $440 million as part of its "next generation of U.S. human spaceflight capabilities" CCiCap program to support human-rated Dragon capsule development. SpaceX also received a $2.6 billion Commercial Crew Transportation Capability (CCtCap) contract in 2014.[99][100]

On March 2, 2019, SpaceX launched its first orbital flight of Dragon 2 (Crew Dragon). This mission, named Crew Dragon Demo-1, was an uncrewed mission to the International Space Station.

The Dragon contained a mannequin named Ripley, which was equipped with multiple sensors to gather data about how a human would feel during the flight. Along with the mannequin was 300 pounds (136 kg) of cargo of food and other supplies.

The successful Demo-2 launch enabled the first commercial orbital human space flight on May 30, 2020. It was crewed with NASA astronauts Doug Hurley and Bob Behnken. Both astronauts focused on conducting tests on the Crew Dragon capsule. Crew Dragon successfully returned to Earth, splashing down in the Gulf of Mexico on August 2, 2020.[101]

SpaceX successes are directly attributable to the vision, leadership, commitment, and yes, finances and high risk-tolerance of its founder and CEO Elon Musk. Born in Pretoria, South Africa in 1971, he moved to Canada at age 17. Soon after earning degrees in physics and economics at the University of Pennsylvania, the young entrepreneur co-founded an early Internet-mapping company called Zip2, a venture which earned him $22 million when Compaq purchased that enterprise in 1999. Musk then invested that sale profit into a new venture, PayPal, netting him more than $100 million when eBay bought that company in 2002. Since then, he became an early investor and CEO of Tesla Motors, SolarCity (a solar panel company), and SpaceX.

Pursuit of Musk's central goal to dramatically lower the cost of access to space applies a hands-on leadership philosophy and vertical development integration strategy which contrasts markedly with the traditional government model. Given that raw materials to create a rocket actually constitute only about three percent of the cost, about 85 percent of the SpaceX launch vehicle is produced in-house. This includes Falcon 9 rocket engines and stages, principal avionics, and all software. Nevertheless, SpaceX still uses more than 3,000 suppliers where outsourcing is most cost-effective.[102 103]

Musk's vertical integration strategy is proving effective to avoid dependence upon monopolistic pricing by outside suppliers who also sell to competitors. SpaceX launch charges are currently reported to be less than $2,500 per pound to low-Earth orbit for its Falcon 9, and less than $1,000 per pound for its Falcon Heavy. This pricing schedule is already placing heavy market pressure on Europe's Arianespace (flying Ariane 5) and International Launch Services flying Russia's Proton rocket.[104 105]

Musk has set his sights on reducing the cost while improving reliability of access to space by a factor of ten, stating:

I believe $500 per pound [to LEO] *or less is very achievable.*[106]

SpaceX has publicly indicated that its reusable technology approach may reduce future Falcon launch prices to a $5-7 million range.

SpaceX's new Falcon Heavy, and now, a new Starship vehicle, is intended to wrest business away from a long-time United Launch Alliance (ULA) monopoly which has depended upon access to Atlas 5 Russian-built RD-180 main engines.

Comprised of three Falcon 9 core stages mounted in a side-by-side configuration, it will enable simultaneous launches of two conventional GEO satellites or a single larger DOD satellite. ULA, a Boeing-Lockheed joint venture established primarily for the DOD market, is likely to replace use of RD-180 engines with a next-generation rocket such as the BE-4 which is being developed by the Blue Origin venture discussed later.

Elon Musk's long-term vision for his company extends far beyond low-Earth orbit. Development of a Mars Colonial Transporter and a super-heavy launch vehicle is a major company priority after its Starship and mated Dragon Crew capsule are in regular use. His dream is not only to develop the capability to human colonization of Mars, but to someday even travel there. He has stated:

I'd like to die on Mars…just not on impact.[107]

Musk told a big audience at the International Astronautical Congress in Guadalajara, Mexico on September 27, 2016 that SpaceX was prepared to send colonists to Mars by 2024, and that a self-sustaining city on Mars could be achieved within 40 to 100 years.

He said:

> *...that his goal was to make Mars seem possible. To make it seem it's something we can do in our lifetimes. That you can go.*

Musk's proposal featured a 400-feet-tall rocket launching from a futuristic-looking version of NASA's Cape Canaveral pad 39A where Apollo astronauts departed from for the Moon. The first-stage boosters of this Interplanetary Transport System, now known as Starship, would launch unmanned spacecraft to low-Earth orbit and then return, followed by propellant tankers that refuel those spacecrafts and return to Earth as well. Each booster would be propelled by 42 of SpaceX's new, more powerful Raptor engines that are now being tested and flown.

Development of the Starship has been garnering a lot of interest from space enthusiasts who keenly follow the happenings in Boca Chica, a little village in South Texas where SpaceX has been pioneering a rapid prototyping approach. Such an initiative has not previously been seen in the history of space development.

Musk has also unveiled plans to develop Boca Chica into "Starbase," a space travel hub.[108]

Each of SpaceX's Mars spacecraft would transport about 100 people to the Red Planet at 62,000 miles per hour, soft land, and enable them to walk on the surface. Tickets would initially be priced at about $500,000, eventually dropping in cost closer to one-third that amount. Spacecraft passenger amenities would include large windows for transit viewing, recreational zero gravity games, and luxurious ocean cruise line-style restaurants.

Seizing upon President Obama's call for NASA to work closely with private partners in advancing Mars exploration, Musk proposed that his $10 billion spacecraft development plan will be funded and operated as a private-public partnership. Science writer Eric Berger suggests that while this might seem a little crazy, it might actually offer NASA a pretty good bargain. His editorial in ARS TECHNICA titled *Why Obama's 'Giant Leap to Mars' is more of a Bunny Hop Right Now* points out that the amount is only about one-third what the agency will have spent to fly the first crewed mission on SLS and Orion between now and 2023.[109]

Northrop Grumman Space Systems' Antares Rocket and Cygnus Spacecraft

Founded in 1982 by David Thompson, Bruce Ferguson and Scott Webster, Orbital Sciences (later named Orbital ATK) helped to create the Orion Launch Abort System that would have helped NASA astronauts escape in event of an emergency involving the now-cancelled Ares 1 rocket. The company is currently competing with SpaceX for ISS cargo resupply services using expendable foreign-supplied rockets and delivery capsules.

On October 17, 2016, just two years after an October 2014 explosion grounded the booster pending a major redesign, Orbital's Cygnus capsule atop a 139-foot-tall Antares rocket delivered more than two tons of food, supplies and experiments to the ISS. That return to flight was the third launch of the upgraded capsule with increased cargo capacity and improved solar panel arrays. NASA has contracted the company for transport services through the middle of the next decade.

The Cygnus freighter is a pressurized service module with avionics, power, and command and control systems that can carry nearly three tons of payloads. Unlike the SpaceX Dragon capsule, each Cygnus, along with its Antares booster, is sacrificed to burn up during atmospheric reentry after each mission.

The Habitation and Logistics Outpost (HALO) module of the planned Lunar Gateway, a space station in orbit around the Moon, is derived from the Cygnus.

Orbital ATK was awarded a $171 million NASA COTS seed money contract in 2006 to support development of an ISS cargo resupply capability. The company was expected to invest an additional $150 million, with that match split between $130 million to purchase Taurus II/Antares launchers, and $20 million for cargo delivery spacecraft. Then, in 2008, Orbital ATK received another $1.9 billion NASA CRS contract to provide 8 cargo missions to the ISS out of Wallops Island in Virginia. A Cygnus carrier successfully connected to the International Space Station in January 2014, marking a success for the company's first contracted mission with NASA.[110]

The original Antares 100-series and Antares 200 first stage rockets were powered by two engines built in Ukraine by the Yuzhnoye Design Bureau in the Soviet Union during the late 1960s and early 1970s. These were purchased and refurbished with modifications by Aerojet in the 1990s. The series was retired following a catastrophic engine failure during Stennis Space Center testing in May 2014 and the launch failure which destroyed the vehicle and payload in October of that year.[111]

The new Antares 200-series using a Russian-made RD-181 engine enables Northrop Grumman Space Systems to re-compete for small-to-medium missions (up to 15,000 pounds / 6800 km to LEO). The Antares second stage applies a solid fuel rocket motor previously used as a Minotaur-C first stage.[112 113]

Boeing's CST-100 Starliner Capsule

In 2014, Boeing and SpaceX received NASA Crew Transportation Capability (CCtCap) program awards (Boeing received $4.2 billion and SpaceX received $2.6 billion), to become NASA's astronaut space taxis for the ISS. Subject to safety certification, each company was to launch between 2-6 missions with up to 7 passengers. Included were at least one crewed flight test with at least one NASA astronaut aboard plus at least two, and as many as six, crewed ISS missions. The agreement allowed Boeing to sell seats for space tourists (one seat per flight) at a price that is competitive with what the Russian program charges tourists.[114]

Dubbed the CST-100 Starliner, the CST stands for Crew Space Transportation; the number 100 stands for the height of the Karman line which defines the boundary of space…an altitude of 100 km (62 miles) above sea level; and the name Starliner follows Boeing's convention for naming airplanes such as the 787 Dreamliner.

The Starliner is similar to an Orion being built for NASA by Lockheed Martin. With a diameter of 15 feet (4.56 m), it is slightly larger than the Apollo command module but smaller than Orion. Design requirements include an ability to remain on-orbit for up to seven months, be reusable for up to ten missions, and be compatible with multiple launch vehicles, including the Atlas V, Delta IV, Falcon 9, and a Vulcan launch vehicle being developed by ULA. They are being built at the Kennedy Space Center facility in Florida.[115]

Whereas Boeing's CST-100 previously received $18 million for preliminary development 1st phase, $92.3 million for 2nd phase, and $460 million under a CCiCap program, it was dropped from consideration for a multibillion-dollar Commercial Resupply Services cargo award in 2015. Accordingly, its Starliner was being developed strictly as a crew carrier in collaboration with new generation space habitat developer Bigelow Aerospace.[116 117]

The Starliner failed its first uncrewed test flight in December 2019. Due to a problem with the spaceship's computer, Starliner didn't fire its engines at the correct time, missed its rendezvous with ISS, and was forced to return to Earth, unsuccessful.[118]

SpaceDev/Sierra Nevada's Dream Chaser Orbital Cargo Plane

SpaceDev, a wholly owned subsidiary of Sierra Nevada Corporation (SNC) which acquired it for $38 million in 2008, originally proposed its Dream Chaser space plane to launch up to seven people and/or cargo on a human-rated Atlas V 412 rocket. The spacecraft design is based on the HL-20 lifting body tested by NASA, a concept dating back over more than a half-century to the 1957 Boeing X-20 Dyna-Soar which enables the vehicle to launch and land on a conventional commercial airport runway.

NASA awarded SNC $20 million to support Dream Chaser development in 2010, another $80 million in 2011, and $212.5 million more in 2012. NASA dropped SNC's Dream Chaser for additional funding through its Commercial Crew Transport Capability (CCtCap) program in 2014 due to lack of maturity, losing in competitions with Boeing and SpaceX who each received awards. Although SNC protested the decision, the U.S. Government Accounting Office sided with NASA.[119]

In 2016, NASA awarded SNC a contract to provide a minimum of six ISS resupply missions through the year 2024 as part of its Commercial Resupply Services program along with previous awardees Orbital Sciences (now Northrop Grumman Space Systems) and SpaceX. Unlike the passenger version, which was designed for launch outside an enclosure, the cargo Dream Chaser version will feature foldable wings to fit within either a European Ariane 5 or Atlas V rocket fairing.

Dream Chaser's propulsion in orbit is provided by twin hybrid rocket engines fueled with hydroxyl-terminated polybutadiene (HTPB) and nitrous oxide (known more simply as non-toxic and easily stored rubber and laughing gas). The engines can be repeatedly stopped and restarted. The control thrusters used for maneuvering burn non-explosive and non-toxic ethanol-based fuel. The vehicle's thermal protection system uses a tile system which can be replaced in a large group following several flights rather than on a tile-by-tile basis as was done on the Space Shuttle.[120]

SNC's Hybrid Rocket Motor, fueled by HTBD, was used to power Virgin Galactic's first-ever suborbital SpaceShip Two flight vehicle, was designed by Burt Rutan's Scaled Composites and is discussed later.

Bigelow Aerospace's Expandable Modules

Billionaire Budget Suites of America hotel chain owner Robert Bigelow previously demonstrated plans to provide commercial real estate in orbit and beyond and has reportedly already invested more than $250 million towards that goal.

Bigelow Aerospace has since discontinued operations.

As with SpaceX's Elon Musk, Robert Bigelow is a successful self-made, and high-risk-tolerant entrepreneur. Beginning with $14,000 remaining from an original $20,000 loan, he purchased, cleaned up and leased out a house with four apartments behind it in 1968. Within two years he amassed $1 million in assets. By 30 years later, Bigelow was a billionaire who had built an estimated 15,000 short-term-let apartment property units and had purchased another 8,000.[121]

Bigelow's proposed for-lease space construction business centered upon development of expandable modular soft-walled structures that are compactly folded around metal cores for launch and deployed by inflation upon reaching orbit. The outer enclosures are comprised of a pressure-tight bladder surrounded by a cocoon, built from layers of plastic foam and bullet-resistant Vectran fabric (a double-strength variant of Kevlar) up to 40 inches thick in some places. The central metal cores contain electronics and equipment.[122]

Basic technology for Bigelow's expandable modules was developed by NASA for a TransHab project which was cancelled in 2002. His company licensed it in exchange for a $400,000 fee and a commitment for a far more significant development program which reportedly involved tens of millions of dollars.

Bigelow Aerospace launched two small-scale expandable module test prototypes to LEO aboard repurposed Russian SS-18 Dnepr ICBMs. Genesis I was deployed in July 2006, and Genesis II in June 2007.

NASA awarded the company a $17.8 million contract in December 2012 to develop and test a larger 565 cubic foot (16 cubic meter) Bigelow Expandable Activity Module (BEAM). BEAM was launched on a SpaceX Falcon 9 rocket and attached to an ISS docking port to determine how well it holds pressure and provides protection from space radiation. The experimental structure will be jettisoned to burn up in the atmosphere after the experiment is over.[123]

Bigelow had planned to market individual and clustered commercial space habitats of substantial size. A BA 330 series (referring to 330 cubic meter/ 11,700 cubic foot volume) was to be about 45 feet (13.7 m) long, 22 feet (6.7 m) in diameter, and weigh about 43,000 pounds (19,500 kg). A pricing scheme shown on a 2015 Bigelow Aerospace website suggested that a future $51.25 million rental fee would enable a customer to experience 60 days aboard a BA 330, including all transport, training, and food. It advertised:

Bring your clothes and your money. We provide everything else.

A much larger, 2,100 cubic meter/74,000 cubic feet Olympus contemplated for future use in LEO and beyond would have nearly twice the pressurized volume capacity the ISS. With a deployed diameter of approximately 41 feet (12.5 m) and weight of 79-90 tons, it would require a super-heavy launch vehicle for delivery to orbit.[124]

Rocket Lab

Founded by New Zealander Peter Beck in 2006, Rocket Lab has risen rapidly through the ranks in the small satellite launch market to employ their Electron rocket to capture a SmallSat and CubeSat market. The Electron is a two-stage launch vehicle which uses their in-house Rutherford engines.

Rocket Lab's rise in prominence has been made possible by engines which use pumps with battery-powered electric motors rather than a more typical gas generator. Their engine is also largely 3-D printed rather than milled.[125][126]

In March 2021, Rocket lab announced they are developing a new rocket vehicle capable of launching humans to LEO. And in keeping with the current market cost reduction trend, they plan to make it partially reusable following SpaceX's example.[127]

Suborbital Startups and Upstarts

In a suborbital flight, the spacecraft does not do a full revolution around the Earth. It merely goes high up, often crossing the Karman line, which define the edge of space, and falls back down.

Emerging space tourism and government-outsourcing technology markets are attracting a new breed of risk-tolerant adventure capitalists who are bringing fresh investment and innovations that are driving rapid advancements. Some of these developments show great promise to not only dramatically reduce the cost of access to space, but in so doing, also avail that far-reaching domain of opportunity to private citizens and enterprises.

Blue Origin's New Shepard Suborbital Rocket and Capsule

On January 23, 2016, Blue Origin, a company created by Amazon founder/CEO Jeff Bezos in 2004, became the first private enterprise in history to launch, land and repeatedly relaunch the same reusable rocket rather than sacrifice it after one flight. The launcher had been recovered from an earlier successful launch and landing that occurred only two months earlier on November 24, 2015 at Blue Origin's West Texas test facility. The vertical take-off and landing rocket was powered by the company's own BE-3 engine, and all three launches carried a New Shepard capsule named in honor of Alan Shepard who piloted America's first suborbital mission.[128]

Demonstrating the company's motto Gradatum Ferociter (Step by Step, Ferociously), Bezos and his team have reported being fascinated—although not surprised—that getting the used rocket ready for a re-launch required very little work. The crews don't even have to remove the engine.

After deploying the New Shepard capsule at an altitude of 329,838 feet and a speed of Mach 3.72—about 2,854 MPH—the rocket glides in a descent back towards the launch pad. Outside drag brakes reduce the vehicle's velocity as fins steer it through high-altitude crosswinds to a location precisely aligned with and 5,000 feet above the landing pad. At that point the engine re-ignites to further slow the booster to 4.4 MPH as the landing gear deploys and the vehicle descends over the remaining 100 feet to vertically touch down on the pad. Then, the New Shepard capsule glides down with parachutes.

Blue Origin began its vertical launch/vertical landing rocket development as a sub-orbital system for commercial tourism with prototype engine and vehicle flights commencing at its modern 260,000 square foot stone and glass facility at Kent, Washington in 2006. Kent is about a 60-minute drive from Amazon headquarters in Seattle and Microsoft Corp. headquarters in Redmond.

The company's largest publicly disclosed setback occurred with the loss of a test vehicle in 2011 at altitude of 45,000 feet due to a range safety problem when ground personnel lost contact and control.

NASA awarded Blue Origin a $3.7 million contract in 2010 and $22 million more in 2011 to develop an astronaut escape system and a composite space capsule prototype for at least three astronauts as part of its commercial crew CCDev program. A successful rocket escape test was conducted in 2012. The company plans to use New Shepard capsule for suborbital tourism and as a microgravity laboratory. Blue Origin is also developing an orbital version nicknamed Very Big Brother, and in addition to supplying engines for its own use, is developing a new heavy-lifting BE-4 engine through a joint venture with United Space Alliance to replace ULA's Atlas V.[129]

In July 2021, Blue Origin first launched four humans on its New Shepard rocket on a sub-orbital flight. In conjunction with the Virgin Galactic flight carrying 6 people to space merely days before, these two companies are signaling the arrival of a space tourism market.

Unlike Elon Musk's SpaceX, which has received billions of dollars of NASA support, Jeff Bezos has financed Blue Origin's ventures largely with his own money. His personal wealth estimated by Fortune magazine in 2012 to be more than $23 billion, began in 1994 by selling books on the Internet out of his garage. As Amazon's founder he was named Time magazine's Person of the Year in 1999 and turned his attention to space in 2000.

Bezos' long-term goal is to do orbital business, dramatically reducing the cost of space flight from tens to millions or even a sub-million-dollar range to enable hundreds of affordable launches every year. He wrote in a blog post:

> *We are building Blue Origin to seed an enduring human presence in space, to help us move beyond this blue planet that is the origin we all know. We are pursuing this vision patiently, step by step. Our fantastic team in Kent* [Washington State], *Van Horn* [Texas] *and Cape Canaveral* [Florida] *is working hard not just to build space vehicles, but to bring closer the day when millions of people can live and work in space.*[130]

On July 20, 2021, Jeff Bezos, along with three other people were launched into space on a suborbital flight on the first manned mission of the New Shepard. Later, on October 13, 2021, Blue Origin followed up by launching William Shatner—who famously played Captain Kirk in Star Trek—on the same 11-minute suborbital trip to space.

Virgin Galactic's SpaceShipTwo and WhiteKnightTwo

On June 21, 2004, SpaceShipOne designed by legendary aircraft innovator Burt Rutan, became the first privately funded spacecraft to make a suborbital flight, reaching an altitude greater than 62 miles (100 km) where the boundary of space is defined to begin. Repeating that accomplishment on October 4, 2004, its developers won the $10 million Ansari X Prize for reaching that altitude twice within a two-week period with the equivalent of three people onboard and with no more than ten percent of the non-fuel spacecraft weight replaced between flights.

Founded by Sir Richard Branson, Virgin Galactic is proceeding with preparations to provide tourism operations following a tragic October 31, 2014 test flight accident that killed the copilot and injured the pilot. Its vehicle, SpaceShipTwo a nearly twice-larger version of SpaceShipOne, is being manufactured by The Spaceship Company (TSC). TSC was originally formed as a joint venture between Virgin Galactic and Burt Rutan's former company, Scaled Composites, and was funded by Microsoft cofounder Paul Allen. Branson bought out Scaled Composites' interest in 2012, making TSC a wholly owned Virgin Galactic subsidiary. Virgin Galactic also owns and operates the WhiteKnightTwo aircraft that transports SpaceShipTwo to its launch altitude.

The October 2014 accident was caused by premature deployment of a feathering mechanism used for controlled descent during atmospheric reentry; it was deployed while the vehicle was still in powered ascent. A National Transportation Safety Board investigation attributed causes to lack of a fail-safe deployment system and pilot error. That event followed a previous fatal July 2007 accident during a SpaceShipOne fueling test explosion at TSC's Mojave facility. Three employees died and three others were injured by flying shrapnel.[131]

Virgin Galactic is proceeding with SpaceShipTwo development. Commercial flights launched by the WhiteKnightTwo aircraft from the new $212 million Spaceport America facility in New Mexico will carry six passengers along with two crewmembers. Delivery of SpaceShipTwo to launch altitude using a conventional runway departure enables the rocket vehicles to clear populated coastal areas prior to firing engines, representing a large range safety advantage over ground-launched vehicles which present more limited launch site options.

As described by the company, SpaceShipTwo was transported beneath WhiteKnightTwo to an altitude of

50,000 feet—then separated and fired its engines for about 70 seconds to coast up to an altitude of 62 miles. Passengers experienced about five minutes of weightlessness with room…

> …to allow for an out-of-seat zero-gravity experience as well as plenty of windows for the amazing views back to Earth.

SpaceShipTwo then turned back and feathered its rudders, turning them up to 90 degrees in order to increase the drag and control the yaw of the spacecraft for better guidance through the atmosphere. The company explained:

> The feather configuration is also highly stable, effectively giving the pilot a hands-free reentry capability, something that has not been possible on spacecraft before, without resorting to computer-controlled fly-by-wire systems.

At 70,000 feet, then with enough surrounding air, the rudders moved back to enable the vehicle to glide and land on a conventional runway.[132]

Since September 2004, about 700 people reportedly signed up for SpaceShipTwo flights at $200,000 per seat. In 2013, the ticket fare was changed to $250,000 until after the first 1,000 people have traveled in order to compensate for inflation from the time Virgin Galactic got started. Although some original prospects asked for refunds, the majority have not. The company also plans to support government and commercial suborbital science missions. Longer-term plans envision point-to-point suborbital passenger transportation services at speeds up to 2,500 mph.[133]

On July 12, 2021, Virgin Galactic successfully launched six people, including Sir Richard Branson, into space on a sub-orbital trajectory aboard SpaceshipTwo named Unity.

Virgin Galactic plans to conduct regular passenger flights beginning in 2022.[134]

Special mentions also warranted to Relativity Space who is developing a mostly 3-D printed rocket, Astra who are conducting tests of a small launch vehicle at the time of writing and Astranis, a well-funded San Francisco start-up who is designing and launching GEO communication satellites.

Part Six: International Developments

POLITICAL AND ECONOMIC developments are shaping independent and collaborative space plans and initiatives. Russia and China are developing joint plans, both in orbit and anticipating future lunar planetary surface operations. India and other countries are joining in robotic efforts to investigate lunar and Martian geological and mineralogical assets/resources for possible habitation, construction and energy/fuel sources.

The U.S. is leading the Artemis Accords with countries like South Korea, Australia, Canada, New Zealand, Italy, Japan, Luxembourg, The UAE, the United Kingdom, and Ukraine. The Artemis Accords aims to have a sustainable and robust presence of the Moon and beyond.

Just as there are opportunities for economic cost and technology sharing benefits, conflicts arise over concerns regarding military and domestic competition between traditional foes and allies. As an example, the U.S. developing the space force which has both domestic and military purposes and programs.

Due to a resurgence of cold war animosities between America, Russia and China, we see increasing governmental and public concerns. In addition, we have other countries that are caught in the middle between various alliances.

Are we venturing forward for the sake of humanity?

If we were a multi planetary species, are we also multinational pioneers?

1. Xi Jinping
   Source: Public Domain
2. Yulu lunar rover
   Source: Sprt98
3. Tianhe core module being tested
   Source: China News Service
4. Tiangong Space Station
   Source: Shujianyang

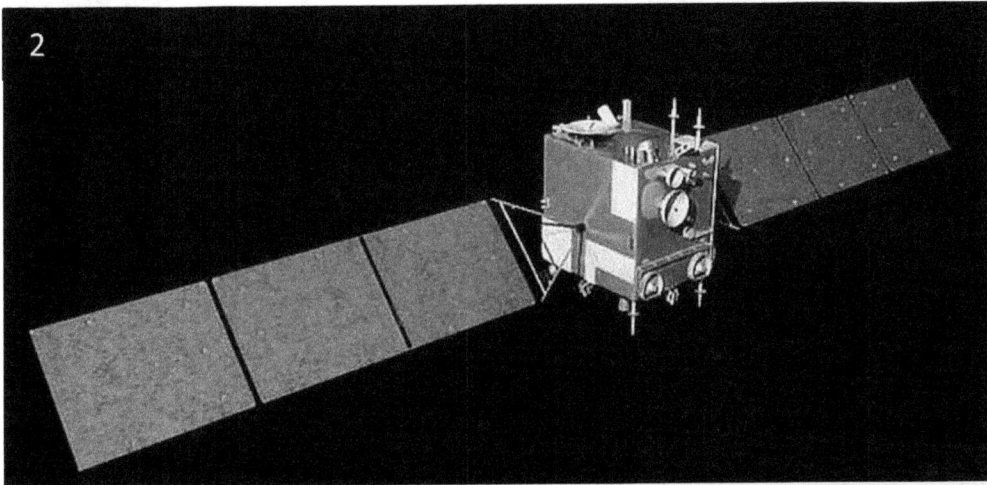

1. Spirit and Opportunity rovers
   Source: NASA

2. Chang'e 2 spacecraft
   Source: CAST

3. Chang'e 4 Lunar lander
   Source: PCAM

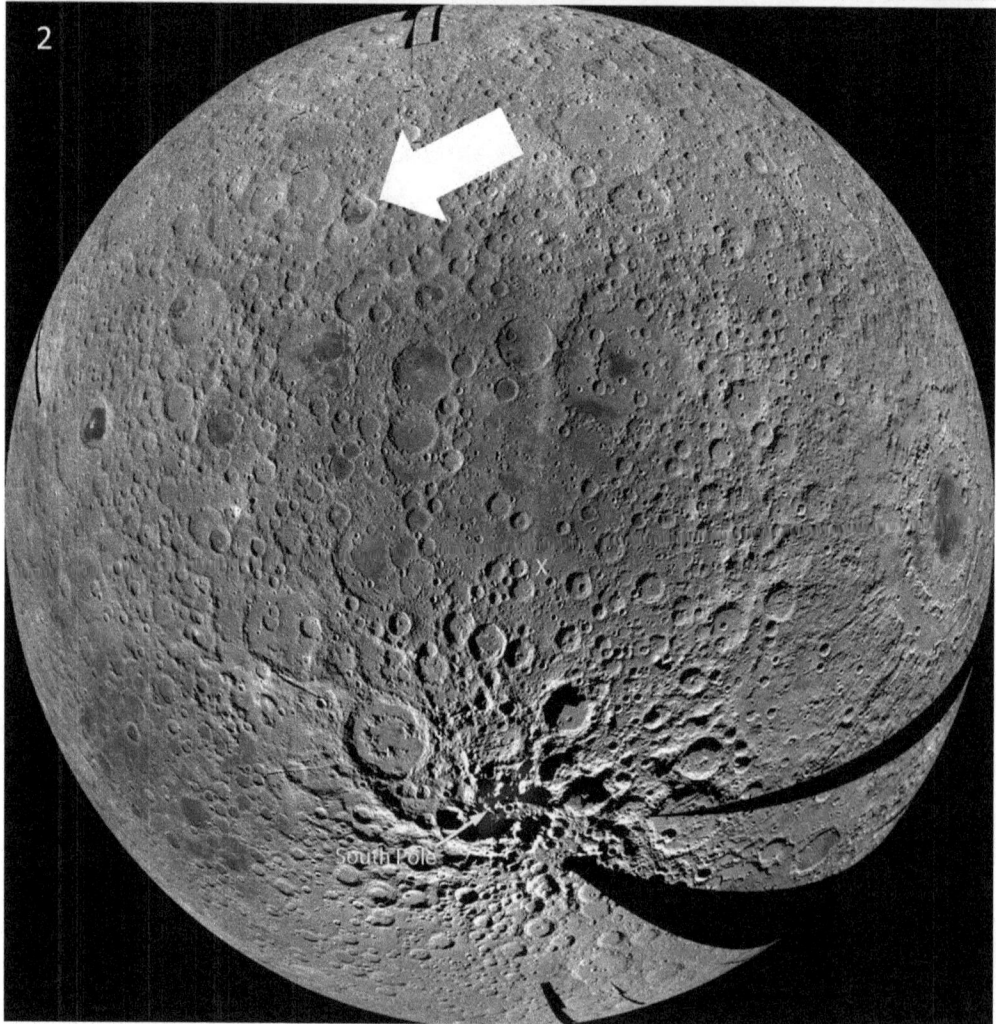

1. The Chang'e-5 Lander and
   Ascender Combination in the
   factory
   Source: China News Service

2. Aitken Basin near the south pole
   Source: NASA

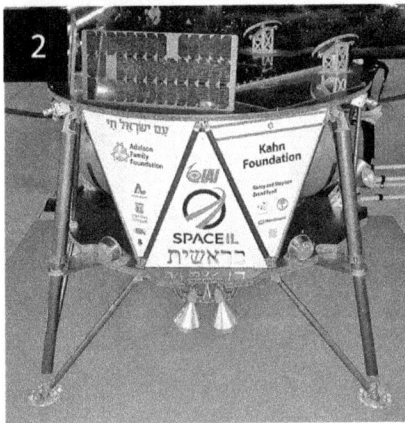

1. Landing of the recovery capsule of the SJ 10 Chinese recoverable satellite.
   Source: Journal of Space Exploration
2. Beresheet lander
   Source: TaBaZzz
3. Vostochny Cosmodrome
   Source: Владислав Ларкин
4. German Herman Oberth is one of the founding fathers of rocketry and astronautics.
   Source: NASA

GPHS-RTG

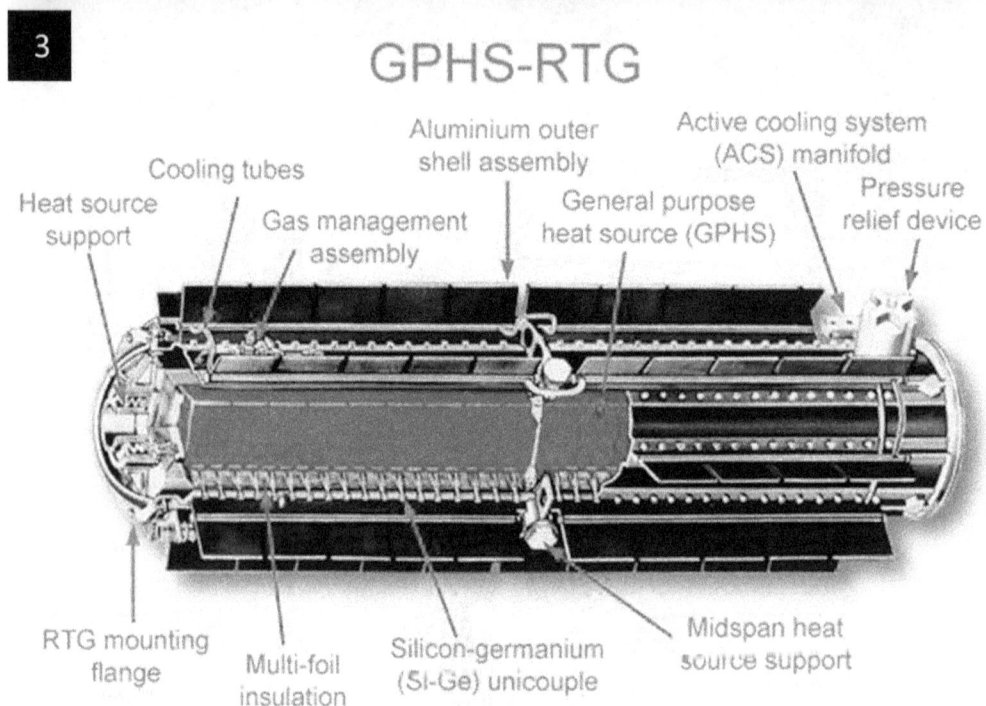

Heat source
support

Cooling tubes

Gas management
assembly

Aluminium outer
shell assembly

General purpose
heat source (GPHS)

Active cooling system
(ACS) manifold

Pressure
relief device

RTG mounting
flange

Multi-foil
insulation

Silicon-germanium
(Si-Ge) unicouple

Midspan heat
source support

1. An ion propulsion engine
   during a hot fire test.
   Source: NASA

2. Solar electric propulsion
   Source: NASA

3. Diagram of the RTG used
   on Cassini probe
   Source: NASA

4. A conceptual spacecraft
   with the VASIMR engine
   Source: NASA

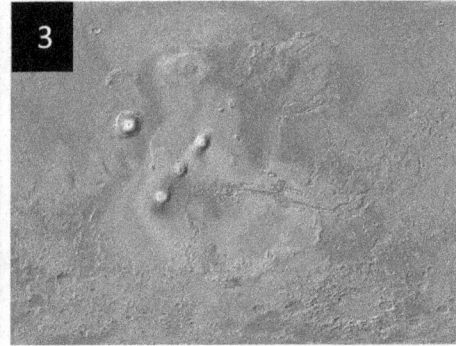

1. Aerial view of *International Thermonuclear Experimental Reactor* (ITER) site in 2020.
   Source: Macskelek

2. Mars Reconnaissance Orbiter reached Mars in 2006.
   Source: NASA

3. The Tharsis region is a vast volcanic plateau of Mars.
   Source: NASA, JP-Caltech, Arizona State University

4. Mariner 9 Probe was launched in 1971.
   Source: NASA

1. Phoenix Lander.
   Source: NASA

2. Viking 1 Orbiter and Lander
   Source: NASA

Chapter Nineteen: Global Coalitions and Competitors

China

PRESIDENT XI JINPING has made it very clear he intends to have China establish itself as a space superpower. China sent its first astronaut into space in 2003, the third country after Russia and the U.S. to achieve independent manned space travel. In June 2016, two Chinese astronauts spent 15 days in orbit and docked with an experimental laboratory, part of Beijing's plan to establish an operational space station by 2020. But that's obviously just a beginning.

On April 29, 2021, Tianhe, the core module of the Tiangong Space Station, was successfully launched into LEO. This is the first module of the long-term Chinese space station, which when completed, will be roughly one-fifth the mass of the International Space Station. At the time of this writing, there is a Chinese astronaut (taikonaut) aboard the Tianhe core module.[135]

China also landed a robotic lunar rover called Yutu (Jade Rabbit), the first to be sent to the Moon in 42 years. The last one was the Soviet's Lunokhod-2, launched in 1973. Designed for operation under both ground and autonomous control, Yutu's tasks include surveying the Moon's geological structure and surface materials while looking for natural resources.

The small six-wheeled 308-pound (140 kg) Yutu contains major design and engineering features that look almost identical to those developed and flown a decade ago by NASA's Mars Exploration Rover program...the Spirit and Opportunity vehicles in particular.

It's important to note that China's lunar lander is far too big to have been designed for tiny rovers. Its size is 40 percent larger than a NASA Apollo module descent stage, suggesting that it must have been engineered from the beginning for the addition of an ascent stage and crew cabin module to carry astronauts. The Chinese are building as many as six such landers on an assembly line basis.

Yutu follows successes of two previous missions in 2007 and 2010. The first spacecraft orbited the Moon for 494 days before departing for deep space. It is now more than 37 million miles (60 million km) away from Earth—China's first man-made asteroid.

The second mission verified some critical technologies and checked out the landing area for Yutu. That small Chang'e-2 spacecraft provided a 23 feet (7 m) high-resolution map of the entire surface of the Moon. Its mapping mission showed the distribution of elements such as uranium, thorium, potassium, and iron on the surface, as well as aluminum. Following completion of its primary lunar tasks, Chang'e-2 was sent out into deep space, and having completed over 125 million miles (200 million km) in flight, will continue to fly, returning closer to the Earth around 2029.

China is reportedly adding new missions to its robotic lunar exploration program to include sending spacecraft to both the north and south poles of the Moon. China's State Administration for Science Technology and Industry for National Defense (SASTIND) told Central China Television that a comprehensive 20-year strategy for lunar and planetary exploration is being developed. The Chinese Lunar Exploration Program (CLEP) will build on the successes of Chang'e-5 near-side sample return mission in 2017 and Chang'e-4 lunar far side mission in 2019.

China successfully launched its Chang'e-5 spacecraft to bring back lunar samples in 2020.

The preservation and handling of extraterrestrial samples requires extremely highly sensitive technology to keep them in a pristine environment and separate from any Earthly contamination. NASA set up a special laboratory at the Johnson Space Center to preserve Apollo program samples to allow for scientific investigations that continue to this day.

SASTIND Chief Engineer Tian Yulong said:

> *The exploration of* [the] *lunar poles is a significant innovation in human history.*

Of particular importance is the south pole Aitkin Basin with the largest crater on the Moon, and one of the largest in the Solar System. Because it is at the pole, there are deep valleys that virtually see no sunlight which are likely to contain water ice.

In 2016, Space officials used the 60th anniversary occasion of China's first satellite launch dedicated as National Space Day to announce future plans. Included were priorities to pursue deep space exploration, continue development of a multi-module space station, and cooperate in international space partnerships. A China National Space Administration (CNSA) display featured a Mars lander scale model.

President Xi Jinping instructed scientists and engineers to:

> *...seize the strategic opportunity and keep investigating to make a greater contribution to the country's overall growth and the welfare of mankind.*

He continued,

> *In establishing Space Day, we are commemorating history, passing on the spirit, and galvanizing popular enthusiasm for science, exploration of the unknown, and innovation, particularly among young people.*

CSNA Director Xu Dazhe said that although China has sent its lunar spacecraft into space:

> *Only by completing this Mars probe mission can China say it has embarked on the exploration of deep space in the true sense.*

The mission includes an orbiter, lander and rover which will walk on Mars.

Experiments aboard China's SJ-10 capsule, launched in April 2016, indicate that national leadership clearly envisions space colonization. Included are mouse embryos to determine if early-stage mammals can develop in extraterrestrial environments. Of the 6,000 total, 600 photographed by a high-resolution camera every four hours for four days successfully developed from the two-cell stage to the blastocyst stage, where noticeable cell differentiation occurs.

Duane Enukui of the China Academy of Sciences told China Daily:

> *The human race may still have a long way to go before we can colonize space. But before that, we have to figure out whether it is possible for us to survive and reproduce in the outer space environment, like we do on Earth.*

He concluded,

> *Now, we finally proved that the most crucial step in our reproduction—the early embryo development—is possible in outer space.*[136]

In 2011, contemplating that U.S. policy prohibiting bilateral space cooperation with China will change, Xu said:

> *When I saw the US film, 'The Martian', which envisages China-US cooperation on a Mars mission under emergency circumstances, it shows that our US counterparts very much hope to cooperate.*[137]

Liang Xiaohong, deputy director of China's Academy of Launch Vehicle Technology has reported that China is beginning development of a Saturn V-class Moon rocket with 11 million pounds (about 50 million N) of liftoff thrust. This is 3.5 million pounds (about 15 million N) more than the Apollo Saturn V. It will be designated the Long March 9.

China's state rocket company is developing a spacecraft which looks a great deal like the Starship from SpaceX, essentially a Starship clone. Recent reports also indicate that China is rethinking their Long March 9 rocket to match the design structure of SpaceX's Falcon Heavy Vehicle. Instead of four booster rockets, the new design will use a single core stage powered by 16 new YF 135 engines clustered in a manner similar to that of the Starship booster stage.[138]

China and Russia have announced a plan to invite international partners to join them in building a Moon base, but they don't plan to send astronauts in the next decade. Officials report they are in negotiations with prospective partners including the European Space Agency (ESA), Thailand, the United Arab Emirates, and Saudi Arabia.

India

Then there's India, which sent a Mars orbiter named Mangalyaan to the red planet with an instrument to measure and map any potential sources of methane plumes which might indicate presence of a microbe biosphere deep beneath the Martian surface.

India has also launched two lunar missions, Chandrayaan-1 and 2. Chandrayaan-1 was put into low lunar orbit in 2008, and later impacted on the surface.[139]

Chandrayaan-2, which included an orbiter, a lander, and a rover, was the first mission to survey the little-explored lunar south pole region.[140]

The lunar orbiter was efficiently positioned in an optimal lunar orbit, extending its expected service time from one year to seven. However, the lander crashed on landing, and India will attempt another soft landing on the Moon with Chandrayaan-3 in 2022, but without the orbiter.[141] [142]

The Indian Space Research Organization (ISRO), having decided not the join the International Space Station project, plans to send humans to space and later aims to establish a space station of their own. They opened a Human Space Flight Center in 2019, in preparation of the Gaganyaan project, a crewed orbital spacecraft.

India has future plans for missions to the Sun, Venus, Mars, asteroids, comets and outer solar system, deploying more telescopes in space and developing satellite navigation systems with global coverage.[143]

Why would India invest in such costly programs?

As Nisha Agrawal, chief executive of Oxfam in India told BBC:

> *India is home to poor people but it's also an emerging economy, it's a middle-income country, it's a member of the G20. What is hard for people to get their head around is that we are home to poverty but also a global power...We are not really one country but two in one. And we need to do both things: contribute to*

global knowledge as well as take care of poor people at home.

K. Radhakrishnan, chair of the Indian Space Research Organization (Isro) said:

> *Why India has to be in the space program is a question that has been asked over the last 50 years. The answer then, and now and in the future will be: 'It is for finding solutions to the problems of man and society.'*

Yes, we have economic challenges, just as our nation did when John Kennedy and the U.S. Congress committed to safely deliver American citizens to the Moon and back within a decade. If India and China now recognize that they can and must make investments necessary to explore the high frontier, why can't we? Can we really afford not to?

Several years ago, Neil Armstrong told BBC science correspondent Pallab Ghosh:

> *The dream remains. The reality has faded a bit, but it will come back in time.*[144]

His prophesy is coming true.

The big question now is if and when it will come back to America.

UAE

The United Arab Emirates Space Agency (UAESA) launched a Mars Hope mission to Mars in July 2020 to create a holistic picture of the chemical composition of Mars' atmosphere. They successfully put the probe around Mars on February 9, 2021.[145]

UAE also has an uncrewed lunar mission in the works to be launched in 2024 when they plan to deploy a rover on the surface of the Moon.[146]

Israel

Israel has had a long history of satellite programs for reconnaissance and commercial activities. They are one of only seven countries that both build their own satellites and launchers.

The Israeli Space Agency (ISA) has been collaborating with NASA regarding future lunar research programs.

In 2019, Space IL, an Israeli private company, supported by the ISA, attempted to place its Beresheet lander on the Moon, a project which hoped to win the Google Lunar X prize. Unfortunately, the lander crashed on the lunar surface. A follow up mission, Beresheet 2, is scheduled to be launched in 2024.[147]

Iran

The Iranian Space Agency has a controversial space program condemned by United States and Europe due to the military potential.

Iran has plans for a manned lunar program scheduled for 2025. According to unofficial Chinese internet sources, an Iranian participation in the future Chinese Space Station program has also been discussed.[148]

Japan

Japan's Japanese Aerospace Exploration Agency (JAXA) has had a good collaboration with NASA and ESA in its space ventures as an International Space Station program partner. Japan is also one of few countries with space launch capabilities, evidenced by their success in a recent asteroid sample return mission named Hayabusa-2. JAXA has active relationships with ESA in planning future robotic missions to Mercury and Venus.[149]

Russia

The Moscow Times reported in January 2015 that Russia is proposing to create a new $150 billion space station for the BRICS group of emerging economies (Brazil, Russia, India, China and South Africa) which would replace the ISS following its originally planned retirement in 2020, but which NASA plans to extend to 2030. Crafted by Russia's Military-Industrial Commission, the government's primary interface with the aerospace and defense industries, the proposal came just months after wide-spread speculation that the crisis in the Ukraine might doom U.S.-Russia space cooperation.[150]

The Russian news agency TASS cited an expert panel on the commission panel stating:

> It would be useful to explore the possibility of an international manned spaceflight project with BRICS nations within an overall strategy of entering into technological alliances.

President Vladimir Putin personally took control of that commission in November of 2014.

The Moscow Times reasoned that a BRICS space station would likely emerge primarily as a two-nation partnership involving either China or India with Russia in the driver's seat. Both countries have well-developed and increasingly ambitious space programs. They, in turn, would encourage the other BRICS nations to invest and participate.

China, which is already launching its own astronauts into space and developing a medium-sized space station, was regarded as the best partner for such a project. The placement of Russia's new Vostochny Cosmodrome in the country's Far East region also facilitates close cooperation with China.

Chapter Twenty: Developing Tomorrow's Technology Today

VENTURING FARTHER, DOING more, and accomplishing economically beneficial and sustainable progress will require and also incentivize technological advancements which must commence now. This will demand investments in new and far more efficient types of rocket propulsion technologies along with long-awaited nuclear energy breakthroughs to overcome limited capabilities afforded by chemical and solar systems.

New Ion Engines for Deep Space Explorations

In reality, the chemical rocket technologies that first opened gateways to a new human frontier of space exploration and development have changed very little since chemical rockets first provided the means to escape Earth's powerful gravitational hold. Yet remarkably, some of the earliest visionary thinkers and experimenters who made this possible, also conceived of a next generation of ion engine technologies that would extend that reach and capacity.

The first documented reference to the concept of ion propulsion is attributed to Konstantin Tsiolkovsky in 1911, and the first discussion of electric ion propulsion was found in Robert Goddard's handwritten notebook entry in 1906. Goddard also conducted experiments at Clark University from 1916-1917 which investigated possible use of ion rocket thrusters under high altitude near-vacuum conditions (although only conceptually demonstrated with ionized air streams at atmospheric pressure). Herman Oberth predicted ion engine application benefits for orbital satellite positioning control and spacecraft propulsion fuel mass savings in his book *Wege zur Raumschiffahrt* (Ways to Spaceflight) published in 1923.[151] [152] [153]

Ion propulsion engines are well suited for very low-thrust applications such as to maintain a satellite's or spacecraft's position in orbit and for non-time-critical transport of satellites, cargo, unmanned habitats and crew return vehicles on deep-space trajectories such as the Mars vicinity. Although capable of attaining very high velocities with low fuel mass, they do so with exceedingly slow but continuous rates of acceleration, beginning gradually like a car with a tiny engine that takes days, weeks or years to reach highway cruise speed, but ultimately reaches thousands of miles per minute on a single tank of gas.

Unlike conventional rocket engines which generate huge, short burst amounts of thrust by a chemical reaction that explosively frees energy contained in the fuel propellant, ion engines provide thrust by ejecting beams of electrically energized atoms or molecules contained in a propellant gas (typically xenon) at extremely high velocities.

The amount of economy relative to the amount of thrust produced by rocket engines (the specific impulse), is a measure of the average exit velocity of gas coming out. Although ion engines provide only tiny amounts of total thrust, nowhere near the amount needed to launch anything from Earth, their specific impulse can be at least an order of magnitude higher…meaning that ten times less in-space transport fuel must be launched to orbit.

Solar Electric Propulsion (SEP)

Ion space propulsion systems in use today add electrical energy provided by external photovoltaic solar arrays to ionize atoms of a propellant gas which is expelled from the back of the engine by a strong magnetic field. Accordingly, the total amount of thrust depends upon how much electrical energy is made available to combine with the propellant gas energy.

For general reference, a single kilowatt (kW) provides only about the amount of power needed to power an electric hair dryer, while over time in space, it is enough to push large things great distances. And while the current state-of-art for SEP systems is only about 5kW, NASA's goal is to eventually boost capabilities up to hundreds of kW for capabilities to rapidly transfer multi-ton payloads to the vicinity of the moon and Mars and back.

Solar electric propulsion (SEP) engines are most commonly used for station keeping to maintain geosynchronous Earth-orbiting communications satellites in their allocated locations. The largest used today aboard large geostationary communication satellites have solar arrays which produce between 20-25 kW.

Three SEP thrusters propelled a Dawn satellite launched in 2007 atop a ULA Delta II booster on a voyage to orbit the giant Vesta asteroid and the dwarf planet Ceres. Reaching a velocity of 22,370 mph (10,000 m/s), Dawn was the first satellite to go around two extraterrestrial bodies. Using chemical propulsion, this feat would have required much heavier spacecraft consuming tremendously more fuel.[154]

A SEP engine also powered a Deep Space 1 satellite mission launched on October 24, 1998, which carried out a flyby of the asteroid Braille. The engine—which produced a thrust equaling about the weight of a single sheet of paper—propelled the satellite to a velocity of 9620 mph (4,300 m/s) while consuming less than 150 pounds (68 kg) of xenon propellant.[155]

The NASA Glenn Research Center in Cleveland, Ohio has been testing a more advanced NASA Evolutionary Xenon Thruster (NEXT) over more than 50,000 continuous hours (equivalent to six years of operation) to test for long-duration missions. It is believed to be capable of generating very high power and thrust levels.

SEP engines are available in two general types: those with gridded thrusters and Hall Effect. Gridded thrusters as used on the Dawn spacecraft to create plasma can thrust chamber which is accelerated by a voltage applied between two semi-permeable screens (like screen doors) that accelerate the charged ions out. The gas (again, typically xenon) has no charge. It becomes ionized when bombarded with energetic electrons provided from a hot cathode filament that fall to an anode and are expelled at high velocity.[156]

NASA Glenn and the NASA Jet Propulsion Laboratory (JPL) in Pasadena, California are investigating a variety of advanced SEP possibilities including a contract to develop a Nested Hall Thruster they awarded to Aerojet Rocketdyne. The two organizations have also been working to create a 50kW SEP robotic spacecraft which would fly to a near-Earth asteroid, collect a 10-16 feet (3-5 m) diameter piece of material, and return it to an orbit around the Moon where astronauts can go and investigate it. The previously proposed Asteroid Redirect Robotic Mission (ARRM) presently has little support.

According to Glenn's senior propulsion engineer Dave Manzella:

> *We could do that mission using only a few thousand kilograms of xenon. Using that, we could bring back something that is 20-30 metric tons.*[157]

A major SEP challenge is to design a large enough solar-electric system to power those engines that can be compacted into a small enough package to fit into the Earth launch vehicle's payload fairing for deployment in orbit. In order to achieve sufficient thrust, that solar system must be able to produce between 600-800 volts.

As Dave Manzella explained:

> *We look at the efficiency with which they stow, and we look at how many kW we can get in a cubic meter, and their mass...how many kW we can provide per kg.*

To meet these demands, NASA Glenn is looking at a new generation of flexible blanket solar arrays of a type used on the ISS. This approach bonds photovoltaic cells to a flexible substrate or mesh type material rather than to rigid panels.

NASA Glenn is investigating two different flexible solar panel manufacturers for AARM. A Megaflex array that was built by ATK Space Systems folds out like a fan. This company also provides a smaller UltraFlex version that was transported to ISS on an Orbital Sciences Cygnus resupply mission. Another company, Deployable Space Systems, Inc., provides a Rollout Solar Array (ROSA) which uses a pair of composite slit booms to roll out an array like rolling out a carpet. Both designs are intended to be capable of producing hundreds of kW power levels.

Major SEP improvements over the past decade demonstrate solar cells with efficiencies over 35 percent, with credible projections up to 40 percent. Several years of experience with high-performance systems are presently capable of generating power levels greater than double those now produced over the same wing area of the 300 kW-class arrays used on the International Space Station.[158]

Since electric thrusters are power-hungry, the ultimate goal and challenge in efficiently delivering large payloads to distant asteroid and planetary vicinities will be to develop extremely large solar technologies which can be placed in LEO with a single launch rather than assembled or constructed in space...or to develop next-generation nuclear space power systems to replace dependence on solar altogether.[159]

Variable-Specific Impulse Magnetoplasma Rockets (VASIMR)

The VASIMR propulsion system is an electromagnetic thruster that uses radio waves to superheat argon or xenon gas propellant. Magnetic fields accelerate and funnel the plasma ions rearward to generate thrust. Sometimes referred to as an Electro-thermal Plasma Thruster or Electro-thermal Magnetoplasma Rocket, the technology can bridge between low-specific-thrust and low-thrust, high specific impulse modes. The latter of these functions in a manner similar to traditional rockets.[160]

While belonging to the same electric propulsion family as SEP, VASIMR differs in the method of ion acceleration by avoiding use of electrodes which can erode over time to shorten operating life. Magnetic shielding to prevent contact of most engine parts with hot plasma enables systems to operate at higher temperatures while also further enhancing long-term durability over other ion/plasma designs. Those temperatures can reach upwards of one million degrees Kelvin...about 173 times hotter than the Sun's surface.[161]

The VASIMR concept was created by former NASA astronaut and Ad Astra Rocket Company chairman/CEO Franklin Chang Diaz who has been advancing its development since 1977. A 2005 VASIMR engine model VX 50 developed by the Advanced Space Propulsion Laboratory (ASPL) at NASA's Lyndon B. Johnson Space Center, in collaboration with the University of Houston, Rice University, the University of Texas at Austin and other institutions, produced 50 kW of total radio frequency (RF) power and 0.5 Newton (0.1 foot-pound-second) of force.[162]

In 2007, Ad Astra announced that the plasma-generation aspect of its VX-100 kW engine's first stage high radio frequency power transmitter had tripled the power of the VX-50 kW. The following year its VX-200 kW model's first stage high frequency DC electricity to thruster power processing efficiency reached up to 98 percent using argon as the propellant. By 2013, the VX-200 had executed more than 19,000 test firings, overall demonstrating greater than 70 percent average thruster efficiency relative to RF power input using argon at full power.[163][164]

NASA awarded Ad Astra a $10 million contract in 2015 to advance development of a VX-200SS engine which can meet a variety of deep space mission needs. For example, a scaled up version or using multiple units of such an engine could be used for proposed tasks which include: powering space tugs capable of rapidly and efficiently transferring many tons of cargo from LEO to lunar, Mars and other orbits; conducting orbital space junk cleanup operations; supporting satellite servicing, repair and repositioning; functioning as a high-powered commercial deep-space catapult to send robotic packages to outer reaches of the Solar System; and deflecting potentially dangerous asteroids as well as capturing and repositioning space rocks for mining and resource recovery.[165]

Accomplishing such capabilities will require parallel development of practical high-capacity next generation nuclear electricity sources to avoid exclusive dependence upon solar collectors. As Franklin Chang-Diaz explains:

> *Part of the problem with electric propulsion...is that it's hard to get enough electricity to power the rocket. Typically, electricity in space comes from sunlight, solar power. That works okay in Earth orbit and other places close to the Sun, but people have to realize sooner or later that, if we're going to explore Mars and beyond, we have to make a commitment to developing high-power electricity sources for space.*

Diaz-Chang emphasized:

> *If we don't develop it, we might as well quit, because we're not going to go very far. Nuclear power is central to any robust and realistic human exploration of space. People don't really talk about this at NASA. Everybody is still avoiding facing this because of widespread anti-nuclear sentiment.*[166]

Nuclear Options for Future Space Operations

The Director General of Roscosmos Vladimir Aleksandrovich Popovkin agrees with Diaz-Chang, saying that development of megawatt-class nuclear space power systems is vital in order for Russia to maintain a competitive international space future which includes human exploration of the Moon and Mars. Accordingly, the government has recently dedicated an equivalent of $540 million USD to begin designing a nuclear lunar/Mars surface station with a 10-15-year service life. Roscosmos is also developing a nuclear-powered space tug concept.[167]

The UN's 71-nation-member Office for Outer Space Affairs (UNOOSA), which implements decisions of a Committee on Peaceful Uses of Outer Space (COPUOS), recognizes that nuclear power will have a legitimate and important role in future space development. UNOOSA's position has stated:

> *That for some missions in outer space nuclear power sources are particularly suited or even essential owing to their compactness, long life and other attributes, and also that the use of nuclear power sources in outer space should focus on those applications which take advantage of the particular properties of nuclear power sources.*

Further, UNOOSA has adopted a set of principles applicable to nuclear power sources in outer space devoted to the generation of electric power on board space objects for non-propulsive purposes, that include various types of radioisotope systems and fission reactors. The former are typically small devices used to provide very limited power levels, while the latter, afford greater potentials.

Traditional Small-Scale Radioisotope Applications

Although the notion of putting any sort of nuclear device in space may seem startling to some people, this has been going on for more than 50 years with little reason for worry. The far most common of these are Radioisotope Thermoelectric Generators (RTGs) used to provide electricity for satellites and/or heat to keep instruments warm.

Having no moving parts, RTGs offer the simplicity of solar power systems along with other important advantages. Included are lower weight-to-capacity ratios and volume-to-capacity ratios with no dependence on access to sunlight. They are sometimes used to provide heating for cold-sensitive instruments, and unlike larger nuclear devices, require minimal shielding to protect radiation-vulnerable equipment.

RTGs have been used to power dozens of different types of small craft, both in orbit and on planetary surfaces.

On the larger end of the RTG power spectrum, the Russian TOPAZ-II reactor uses uranium fuel to produce 10 kW through direct thermoelectric conversion.

Russia Amping up Nuclear Space Power Advancement

Today's technologies rely upon nuclear fission technologies for spacecraft needing more than 100kW, and Russia already has plans for much larger systems. Their government allocated federal funds in 2010 to design a megawatt-class nuclear propulsion unit capable of powering a craft on long-haul interplanetary missions.

A conceptual design at Energia (SP Korolev Rocket and Space Corporation) produced engineering details for 150kW-500kW systems using small gas-cooled fission reactors which produce heat to power electricity-producing turbines. It was reported in 2015 that engineering tests confirmed the integrity of the reactor vessel, and a prototype for space applications was then expected to be completed by 2018. The first launches were envisaged for about 2020.[168]

The U.S. has explored space-applicable nuclear fission technologies leading up to 100kW capacities since the mid-1960s. Unfortunately, there is no evident commitment to bring them into fruition any time soon.

NASA first developed and flight tested a 45kW System for Nuclear Auxiliary Power (SNAP-10A) in 1965. The system operated for 43 days until a voltage regulator malfunction caused a power shut-down. SNAP was followed in 1983 by the joint NASA-DOE-DOD development of a 100kW SP-100 Space Nuclear Reactor Program. Intended to power orbiting missions or to provide lunar/Mars surface power, that project was terminated by GAO in the early 1990s after spending nearly $1 billion:

> ...in the belief that another nuclear reactor technology would more cost-effectively and timely meet its needs.[169]

Future Fission Reactors Fueled by Lunar Thorium?

Future power abundance needed for spacecraft ion propulsion and planetary surface operations may be provided using small, safe and versatile nuclear fission reactors fueled by thorium. Referred to as Liquid Fluoride Thorium Reactors (LFTRs), such systems offer potentials to be scalable to any size needed, with energy density comparable to uranium but producing only a tiny fraction of the radioactive waste. In addition, unlike other fission fuels, the power-producing process can be immediately shut down as desired by simply switching off the photon beam that drives the chain reaction process.

Thorium is a very plentiful material on Earth as well as on the Moon and Mars. It is also very inexpensive, a byproduct of rare earth material mining to supply international electronics industries.

Unfortunately, while LFTR technology isn't particularly new in concept, it still has a long way to go before it demonstrates fully competitive commercial performance and efficiency potential. U.S. research and development demonstrations date back to experiments with a 7.4 MW (thermal energy) molten salt reactor type at the Oak Ridge National Laboratory which ran from 1965 to 1969. The project was shelved by the Richard Nixon administration due to Cold War imperatives. The Pentagon was more interested in using plutonium residue from uranium-fueled reactors to build bombs.[170]

Far more active thorium reactor research and development pursuits are currently underway in other countries, including China, Japan, Norway, and India.

The China Academy of Sciences and Shanghai Institute of Applied Physics (SINAP) launched a significant LFTR program in 2011, with plans to build a small 2 MW plant by the end of decade before ramping up to a commercially scaled plant by the 2020s.[171] [172]

The challenge is to demonstrate so-called fourth-generation molten salt reactors which operate at much higher temperatures and lower pressures than current technologies. Meanwhile, although there is some work being done on third-generation light water reactors that are more efficient and safer, the U.S. continues to rely upon second-generation light-water, solid fuel conventional reactors.

Thor Energy in Norway has built a small test reactor using rods of thorium and plutonium oxide to drive a steam turbine for electricity at a paper mill. The program is being undertaken in collaboration with Japan's Toshiba-Westinghouse. Japan's International Institute for Advanced Studies (IIAS) is pursuing research on thorium-fueled molten salt reactors.[173]

India, a country with limited natural supplies of uranium, has made advanced development of thorium-fueled heavy water fast breeder reactors a major goal to meet growing electricity needs. The nation's time schedule for accomplishing this is vague, most likely requiring at least 15-20 years.[174]

Beyond Fission to Fusion?

Science aimed at extracting nearly unimaginable amounts of energy by fusing tiny amounts of fuel continues to be an elusive quest. Described as somewhat like trying to put a sun in a bottle, it involves spinning a cloud of hydrogen plasma suspended in a vacuum by high-power magnets to a temperature of several hundred million degrees Celsius until the positively charged atoms are stripped of electrons which hold them apart and fuse into helium.

In doing so, the atoms spit out extra neutrons which become embedded into a surrounding blanket of lithium and warm it enough to boil water into steam that spins a turbine to produce electricity.

Experimental fusion programs began in the U.S. in the late 1970s at MIT and Princeton. In fact, Princeton's Tokamak Fusion Test Reactor (TFTR) actually succeeded in producing 10 MW...but only did so for about a second. TFTR was shut down and replaced in 1999 by the university's Plasma Physics Laboratory's National Spherical Torus Experiment, a $94 million 22-foot-tall metal spheroid surrounded by magnets rising vertically from the floor like fingers clutching a ball.[175]

The world's largest nuclear fusion program by far, is the International Thermonuclear Experimental Reactor (ITER) located in Southern France. Costing more than $4 billion, the collaborative scientific effort is backed by the U.S., EU, Soviet Union/Russia, Japan, India, South Korea, and China.

The long-term goal is to create a self-sustaining fusion reaction which produces more energy than needs to be put in. As many scientists observe, that goal was predicted to be 30 years away 30 years ago, and still remains to be 30 years away now. China and South Korea are reportedly more optimistic, hoping to achieve success by the 2040s.[176]

China's research is presently focused on using deuterium and/or tritium (heavy isotopes of hydrogen). While deuterium is abundant in all water on Earth, tritium is not found in nature and is produced through neutron bombardment of lithium. In combination with plans to establish a base on the Moon, the Chinese government is also believed to be interested in exploring another future fusion isotope option on the periodic table. Helium-3, a molecule which is almost non-existent on Earth, is prevalent in lunar soil.

Helium-3 has been deposited by solar winds on the Moon over billions of years to an estimated depth of a few meters. If world fusion technology proves successful, it is estimated that each ton of the material would yield energy equal to approximately 50 million barrels of crude oil. This suggests that about 40 tons of He-3 might potentially power the entire U.S. for a year at present consumption rates.[177]

The relative economy of collecting, processing and transporting that material to Earth remains another major unknown. Mining helium-3 or anything else on the Moon will require the construction and operations of a significant industrial infrastructure along with enough power to extract He-3 by heating lunar dust to 600 degrees Celsius.

We can then only wonder...will Thorium fission or He-3 fusion advancements eventually provide that power using fuels harvested on the Moon?

And if so, will America be the nation that develops them and realizes the benefits?

Chapter Twenty-One: Pioneering New Worlds of Opportunity

EXTENDED DISTANCE AND time operations on human space voyages and lunar/planetary surfaces will require that means are afforded to harvest and use extraterrestrial resources for propulsion, life support and perhaps eventually, for construction and even commercial export. If the latter of these possibilities seem truly remote, consider how leading authorities viewed notions of heavier-than-air flying machines prior to the Wright Brothers' first flight in 1903 and Robert Goddard's predictions less than two decades later that rockets could reach the Moon, much less that humans would walk on its surface little more than a half century later.

Even then, back in 1972, who could have predicted the rocket-satellite-driven Internet systems, miniaturization of computers and smart phones which have, in a virtual sense, shrunk the world and very literally expanded information access? Now, less than a half century later, who can imagine the possibilities awaiting the world a few decades from now, or those that will be foregone if we don't prepare for them now?

Living off the Land

Space affords a natural supply depot stocked with a vast assortment and quantity of potentially useful, even critical, materials that can expand human experience and enterprise. Of these rich resource caches, the Moon is the closest, representing a relatively near-term source and laboratory for rocket propellant, oxygen and water production, and an operational base for development and demonstration of other extraterrestrial technologies.

The Moon has some similarities to Earth, relatively close access to solar energy, some gravity and plentiful resources such as water, oxygen and minerals, for example. There are also dramatic differences. Included are the lack of any atmosphere, extreme temperatures, and the fact that harvesting and processing those resources to support substantial scale operations will impose daunting technological and infrastructure challenges.

Solar circumstances on the Moon create unique conditions affecting virtually all planning considerations. The Moon's slow rotation on its axis (once in about 708 hours) produces one lunar day for slightly less than 29 Earth days, with about 14 days of uninterrupted sunlight. And since the inclination of the Moon's spin axis is only 1.5 degrees as compared with 22.5 for Earth, if you're standing at the poles the Sun circles around you near the horizon instead of rising and setting as we see it.

In fact, if you're standing on a solar peak, you will see it all the time, meaning that a solar electricity generating system located there will produce constantly. Also, because the Sun appears near the horizon, its illumination grazes the surface so there are no big temperature swings to have to adapt to. Instead, nearly permanent sunlit areas remain at a not-so-balmy -58 degrees Fahrenheit (-50 C).

If you imagine that this is cold, temperatures in lunar craters and other shadowed areas are far more extreme…reaching about 25 degrees F above absolute zero. There are estimated to be between 2,300-5,800 square miles of permanently shadowed area around the Moon's South Pole, and that's where lots of the coveted water-rich bounty resides.[178]

Lunar Water

Recent discoveries of large quantities of water on the Moon have excited great interest for a multitude of applications which can greatly reduce dependence upon costly transportation of Earth-delivered consumables. The precious molecule is vital to life for drinking and constituent source of oxygen for breathing; as a source of hydrogen and oxygen for rocket propellant to fuel ascent vehicles from the lunar surface and transportation to cislunar (Moon-Earth-orbit) space and beyond; and can also provide a highly mass-efficient solar and nuclear radiation shielding material.

The Moon's surface is constantly being bombarded by meteorites and cometary nuclei that deposit volatile molecules such as water-ice and water vapor on its surface. Although the weak lunar gravity and lack of atmosphere (and clouds) allows most of the water vapor to rapidly escape back into space, ice particles and gas vapor that enters colder shadowed areas often become trapped.

It is theorized that some water may also be released from deeper towards the Moon's interior, suggesting a previously "wet Moon." This possibility is suggested by magmatic water found in Apollo mission volcanic glass samples. Recent re-analyses have found the contents to be very similar to those of primitive terrestrial mid-ocean ridge basalts on Earth, indicating that some parts of the lunar interior may even contain as much water as Earth's upper mantle. This line of reasoning follows a possibility that the Moon was formed from a giant impact collision between a roughly Mars-sized object and the proto-Earth.[179]

The existence of water on the Moon is now a certainty. NASA's Lunar Prospector satellite launched in 1998 carried a neutron spectrometer that detected hydrogen at both poles, indicating a large quantity of water ice mixed with lunar soil (regolith).

Mini-RF imaging radar aboard an Indian Chandrrayaan-1 spacecraft launched in 2008 found spectral evidence in an extremely thin layer of water molecules at highest latitudes across all types of lunar terrain. It appeared that the water, likely derived from solar wind deposits, was in motion, driven by local sunlight heating towards colder locations. Chandrrayaan-1 also deployed an impact probe into a dark shadowed region above the Moon's South Pole which threw a plume of water vapor and ice crystals into space.

The amount of water contained in cold shadowed areas is believed to be immense. NASA's Lunar Cratering Observation and Sensing Satellite (LCROSS) impact probe which were launched at the same time (2009) found nearly pure water deposits several meters thick within a wall of the South Pole region Cabeus crater, suggesting that tens of billions of metric tons of water might be accessible for use.[180]

NASA is tentatively planning two robotic missions to take a look at locations where water is near the surface. The Lunar Flashlight mission involves a tiny satellite about the size of a cereal box which will unfurl a large 860-square-foot solar sail that will propel it to lunar orbit by capturing photons streaming from the Sun. Reaching the Moon six months after launch, it will spend a year spiraling down to an altitude about 12 miles above the surface to bounce sunlight off its sails like a mirror and measure and map water ice deposits near the poles.

As Lunar Flashlight inventories water from above, a planned Resource Prospector Mission (RPM) rover will take a closer look at two promising surface and subsurface locations about 1 km apart. RPM will be equipped with a drill to take samples up to 3.3 feet deep which will then be heated by an onboard oven and instruments to liberate, identify and quantify water and other volatile contents. The solar-powered rover and its equipment will rely upon batteries with an operational lifetime of about one Earth week upon entering shadowed areas.[181]

Lunar Prospectors and Prospects

In addition to water, the Moon's surface is known to contain a vast variety of other resources to help sustain mission operations and potentially afford commercial opportunities for a new breed of adventure capitalists. For example, lunar soil (regolith) contains abundant amounts of chemically bound oxygen for life support and propellant; metals including aluminum and iron; basalt which can be cast into ceramic bricks and sintered for paved surfaces; silicate to produce photovoltaic cells; and valuable rare earth elements (most notably helium-3).

Accessing, separating, and processing those substances, on the other hand, presents enormous challenges. In addition to the fact that enormous costs of delivering cargo to the Moon will stringently limit equipment sizes, inventories and maintenance spares, harsh and unique environmental conditions will impose a variety of operational difficulties. All such operations and maintenance will need to be highly automated with little or no human hands-on intervention when something inevitably goes wrong.

Airless vacuum and low gravity (1/6th Earth gravity) conditions will impact many operations and systems related to digging, separating, processing, containing, and transporting resource materials. The vacuum condition causes out-gassing from the soil which produces more friction by super-tiny and abrasive surface dust particles and will rapidly degrade drilling, moving mechanical parts, and critical pressure seals. Huge temperatures variations ranging from about -275° F (night) to about +280° F (day) and differences between sunlit and shadowed areas will influence lubricant viscosity and battery efficiency. Long day/night diurnal cycles will impact power requirements and battery storage needs.

Industrial-scale excavation and processing of in-situ resource materials will impose large power demands. Providing that this involves extraction of water which appears most plentiful near the poles, some or all might be obtained from solar panels positioned at a high point where direct sunlight is available virtually all the time. In this instance, the water might be obtained from a nearby shaded region through a sublimation process that heats and collects it as a gaseous vapor.

Surface oxygen and various rare earth elements might preferably be mined from polar locations situated where sunlight continually grazes the surface and temperatures remain constant. Large fields of photovoltaic arrays to collect this energy could theoretically be processed from silicate at the same locations.

Particular mining and processing methods will depend to some degree upon the particular site and material that is to be harvested and the cost vs. value the seekers are willing to invest. The top few meters over most of the Moon's surface are thought to consist of a large mix of minerals derived from splashes from asteroid impacts, while lower levels are likely to have more uniform minerology originating from old magma oceans. Whereas the top surface tends to be glassier than lower layers due to the superheated nature and rapid cooling of asteroid ejecta, Apollo missions demonstrated that extraction of subsurface samples was actually more difficult due to greater compacted density which caused drilling tubes to seize.[182]

Whether or not the value of mining for lunar resources exceeds high costs and risks involved will depend upon two yet-to-be-determined, separate but interdependent futures. One will hinge upon long-term national and international space exploration and development commitments where necessary investments can sustainably reduce transportation fueling and human life support costs. The second will require government and industry sponsored innovations to dramatically drive down those costs to create high-value commercial market incentives.

Referring to the Obama administration's cancellation of the Constellation program which was intended to return Americans back to the Moon by 2020, former George Washington University's Space Policy Institute Director John Logsdon envisioned private industry partnerships with government as:

> ...a way of NASA getting back involved with the Moon without violating the president's policy that says as a government we don't go back to the Moon.[183]

Space resource entrepreneurs face legal as well as technical challenges which will determine whether prospective enterprises will ever get off the ground. Key among these impediments are regulatory uncertainties regarding private companies' ability to secure essential extraterrestrial land use and mining rights.

The 1967 UN Outer Space Treaty governs what countries can and can't do on the Moon but leaves private companies unregulated. Lines of separation between government and private rights are particularly unclear for countries like China where large companies are state-owned. For example, before China's Yulu Jade Rabbit Moon rover malfunctioned, their public media had reported that its bottom-mounted ground-screening radar was searching for valuable minerals.[184]

Current Outer Space Treaty prohibitions, which prevent nations from claiming territorial rights on the Moon, are widely interpreted as also precluding private companies from owning lunar resources. Ian Crawford, a professor in the Department of Earth and Planetary Sciences at the University of London's Birbeck College urges that international understandings should be updated to provide a legal framework to define and protect corporate interests. He argues:

> *Just as no nation-state can currently appropriate the Moon there is a case for ensuring that private companies also cannot claim to own the Moon, but nevertheless would be legally entitled to materials that they extracted from it as a result of their private investment.*[185]

The U.S. Congress passed a law in 2015 which supports rights of commercial space ventures to extract, use and sell resources from the Moon, asteroids and other celestial bodies, consistent with international obligations. Presently, the only formal requirement is for companies to obtain blessings and supervision for their operations from their host nation which will be responsible for anything that goes wrong. This is based heavily upon a framework of good faith understanding between space-faring nations.[186]

The Google Lunar X Prize

The Google Lunar X Prize was a 2007-2018 program which planned to award a $20 million grand prize to the first team to land a robot that successfully travels 1,640 feet and transmits back high-definition images and video from the Moon. The second to do so would receive $5 million. The qualifying contestants could then win additional money by completing tasks extending beyond baseline requirements such as traveling ten times farther, capturing images of remains of Apollo program hardware or other man-made objects on the Moon, verifying recently discovered surface water ice, or surviving a lunar night.

On 23 January 2018, the X Prize Foundation announced that "no team would be able to make a launch attempt to reach the Moon by the [31 March 2018] deadline…and the US$30 million Google Lunar XPRIZE will go unclaimed." [187]

Along with SpaceIL, an Israeli non-profit company, Mountain View, California-based Moon Express is one of two private companies who reached the final stages to land the first non-government probes on the Moon, in connection with $30 million in prizes arranged in 2007 by Google co-founders Larry Page and Sergey Brin.

Moon Express was awarded a $1 million milestone prize for being the first to successfully test its coffee table-size prototype lander that is designed to softly deliver its probe to the surface. The company has signed an agreement with NASA to take over Space Launch Complex 36 at Cape Canaveral for continued development and testing.

As originally planned, the $20 million first prize would decrease to $15 million in the event that a government-led mission landed and explored the surface before the competitor accomplished this. That rule was dropped after the Chinese Chang'e 3 probe landed in November 2013.

In addition, an original 2015 deadline was extended to December 2017 provided that at least one team could secure a verified launch by December 31, 2015. Moon Express and SpaceIL both achieved this requirement with Rocket Lab and SpaceX, respectively. Structured as a U.S. company with a New Zealand subsidiary, Moon Express will launch its probe and lander from a pad on New Zealand's Mahai Peninsula.

On 11 April 2019, the SpaceIL spacecraft crashed while attempting to land on the Moon. The SpaceIL team was awarded a $1 million Moonshot Award by the X Prize Foundation in recognition of touching the surface of the Moon.[188]

Destination Mars

While the Moon's relatively close proximity to Earth, surface water for hydrogen and oxygen rocket propellant and human life support, and precious rare materials such as He-3 make it a logical near-term supply base and technology laboratory for future exploration beyond, the more Earth-like characteristics and greater resource variety afforded by Mars make it far more attractive as a true destination.

Of special importance, recent findings obtained from NASA's Mars Reconnaissance Orbiter (MRO) offer evidence that as with the Moon, water also exists on Mars, and that some even intermittently flows on its surface. MRO's imaging spectrometer detected darkish streaks up to a few hundred meters in length along with spectral signatures of hydrated minerals which darken and extend down deep slopes during warmer seasons (above minus 10 degrees Fahrenheit / -23 C). These features fade and eventually disappear altogether in cooler seasons.

Although Mars temperatures are characteristically very cold, liquid water is unstable over much of the planet due to low atmospheric pressure. It is theorized that hydrated salts lower the freezing point of the liquid brine, just as salt causes ice and snow on roads here on Earth to melt more rapidly. It's presently unclear whether the dark streaks are signatures of the salts, or rather, appear as a result of the existence of briny water that periodically wicks up from subsurface sources.

Lujendra Ojha, a lead author of a Nature Geoscience report on these findings, believes that the spectral signatures produced by the salts are consistent with a likely mixture of various percholates, which have been shown capable of keeping liquids from freezing under conditions as cold as minus 94 degrees Fahrenheit (-70 C). Of notable interest, some types of naturally produced percholates found on Earth deserts can be used as rocket propellants.[189]

As John Grunsfeld, an astronaut and associate administrator of NASA's Science Mission Directorate in Washington observed in 2015:

> Our quest on Mars has been to 'follow the water,' in our search for life in the Universe, and now we have convincing science that validates what we've long suspected.

He continued:

> This is a significant development, as it appears to confirm that water—albeit briny—is flowing today on the surface of Mars.

Michael Meyer, lead scientist for Mars Exploration at NASA Headquarters, commented on the significance of this discovery stating:

> It seems that the more we study Mars, the more we learn how life could be supported and where there are resources to support life in the future.[190]

Mars has a great diversity of other resources, some of which are present but most likely less accessible on the Moon. For example, carbon dioxide, nitrogen and hydrogen exist on the Moon, but unlike Mars, only in tiny parts per million quantities. Mars isn't known to have helium-3, but does have an abundance of deuterium, a heavy isotope of hydrogen for future nuclear fusion. And while oxygen is abundant in lunar soil, it is tightly bound in oxides such as silicon dioxide (SIO_2), magnesium oxide (MGO), ferrous oxide (Fe_2O_3) and aluminum oxide Al_2O_3) which require high energy processes to release.

As on Earth, hydrologic and volcanic processes on Mars are likely to have consolidated various useful elements into concentrations of high-grade ore. Groups of meteorites originating from Mars reveal magnesium, aluminum, titanium, iron, chromium and trace amounts of lithium, cobalt, nickel, copper, zinc, niobium, molybdenum, lanthanum, europium, tungsten and gold. Mars landers Viking I, Viking II, Pathfinder, the Opportunity rover, and the Spirit rover found aluminum, iron, magnesium and titanium.

The Spirit and Opportunity rovers both found nickel-iron meteorites sitting on the surface, materials used to produce steel.[191] [192] [193]

Dark sand dunes believed to be caused by volcanic basalt rock which are common to the surface of Mars are thought to contain chromite, magnetite and ilmenite that might be used to produce chromium, iron and titanium. Other large areas contain troughs, called fossa stretching thousands of miles out from volcanos that reveal observable dikes formed by underground lava flows believed to have transported minerals including nickel, copper, and platinum. The Tharsis Bulge, a region in the southern hemisphere about the size of North America, contains a group of such giant volcanoes.[194] [195]

The Martian environment has similarities to Earth's, but also some very large differences. Like Earth, and unlike the airless Moon, Mars has an atmosphere, mostly carbon dioxide, although not nearly as dense. Its atmospheric pressure is about equal to the Earth's atmospheric pressure at 100,000 feet. Also, possessing about 11 percent of Earth's mass, Mars has approximately one-third as much gravity…about twice that of the Moon.

There is evidence that the Martian atmosphere was once even denser than Earth's, whereby the constituent CO_2 was gradually removed over eons of time by solar winds. It is theorized that this scrubbing process continues to occur as ultraviolet light from the Sun first breaks CO_2 molecules into carbon monoxide and oxygen; then a second photon of ultraviolet light breaks the carbon monoxide into carbon and oxygen which releases enough energy for the light isotope of carbon (C^{12}) to escape the planet.[196]

The environments of Mars, Earth and the Moon each have important similarities and differences.

Like Earth, with an atmosphere and an axial tilt (25.2°), Mars has seasons and weather which varies somewhat from year-to-year. Although Mars days are nearly the same length of time as on Earth (24 hours and 37 minutes), each Mars year equals 687 Earth days, about two Earth-years. Mars is about 50% farther from the Sun than the Earth and Moon are, and also different, the airless Moon doesn't filter away surface sunlight as the planetary atmospheres do. Massive planet-wide dust storms on Mars present an extreme case. Yet while on one hand these features afford some lunar surface solar energy benefits, the Moon's 28-Earth-day light-dark cycle counteracts these advantages.

Mars is dryer and colder than Earth, and dust raised by winds remains in the atmosphere with no rain to wash it out except for occasional CO_2 snowfall. A 1971 Mariner 9 probe observed a near planet-wide dust storm which lasted a month, an occurrence now found to be quite common. Wind speeds can increase rapidly, and the low density and low gravity can cause dust to remain in the atmosphere for extended periods. In mid-2007, a planet-wide dust storm presented a serious threat to the solar-powered Spirit and Opportunity Mars exploration rovers, necessitating the shut-down of most science experiments until it cleared.[197]

Dust storms kick up fine particles in the atmosphere causing very faint clouds to form very high up…up to 62 miles. A Phoenix lander took pictures of snow falling from clouds above its landing site which vaporized before reaching the ground (a phenomenon called virga).[198] [199]

Like Earth, Mars offers true climate variety. Surface temperatures average about -67°F, can reach a high of +68°F at the equator (noon), and drop to a low of -243°F at the poles. The warmest soil estimated by the Viking Orbiter was 81°F. The Spirit rover recorded a maximum daytime air temperature in the shade of 95°F, and regularly recorded temperatures well above 32°F except during winter.[200] [201] [202]

There is a significant dichotomy of climates between the northern and southern hemispheres during the Martian annual cycle: a northern spring and summer which is relatively cool, not very dusty, and relatively rich in water vapor and ice clouds; and a southern summer with warmer air temperatures, less water vapor and higher atmospheric dust levels.

Southern hemisphere winters are longer and colder than those in the northern hemisphere.

Martian Moons as Strategic Mission Resources?

Mars has two small moons, Phobos (fear) and Deimos (panic), named after horses that pulled the chariot of the Greek war god Ares (Mars). Discovered in 1877 by Asaph Hall, both of these offshore islands are tidally locked to the planet as useful locations to conduct early preparations for human Mars surface missions. Extremely weak gravity conditions will make landing and return to orbit much easier than from the Martian surface, while close proximity can enable remotely controlled robotic habitat development without any Earth-Mars time delay.

Landing on Phobos and Deimos would likely apply the same types of equipment systems designed to engage an asteroid. In fact, there is a good possibility that they were originally formed from carbonaceous chondrite asteroids captured by Mars' gravity. Alternative scientific theories hold that the moons were either created by fragments released from asteroid collisions with the planet, or that they were formed earlier out of the same birth cloud as Mars. In any case, both are heavily cratered by later asteroid strikes.

Phobos, by far the larger of the two, has the appearance of a big rubble pile about 14 miles across covered by dust estimated to be about three feet deep. Although about eight times more massive than Deimos, a still very small size and density provides an escape velocity of only 20 miles per hour. A 150-pound (68 kg) person on Earth would weigh only two ounces there.[203]

Orbiting approximately 3700 miles above the surface, it travels around the planet three times per Martian day. As viewed from the surface, it would appear to be about the size we see Earth's full Moon, perpetually presenting only one side. Traveling an equatorial plane, it wouldn't be visible from Mars' polar caps because it would be beyond the horizon. Surface temperatures on the moon vary greatly depending upon location, ranging from about plus 25 degrees Fahrenheit (-3 C) on the perpetual day side to as much as minus 170 degrees Fahrenheit (-112 C) on the dark side.

Deimos, an irregular potato-shaped body, is only 8 miles across, making it one of the smallest moons in the Solar System. It is so tiny that an object launched from its surface would reach escape velocity at 13 miles per hour. While most debris released by an object impacting the surface will be ejected into space, the gravity from Mars will tend to retain it in a ring in the same orbital region, to eventually be redeposited back on its surface. As a result, a covering of pulverized meteoroid dust estimated to be as much as 150 feet thick partially fills in craters, giving Deimos a smooth appearance from space.

Viewed from Mars, Deimos would seem star like, much like Venus appears from Earth. It orbits around the planet in an equatorial plane in about 30 hours, little more than a Martian day. Being slightly outside synchronous orbit at an altitude of about 12,500 miles above the surface, it takes 2.7 Martian days from rising in the east to set in the west, as it slowly falls behind the planet's rotation.

Following a long history of scientific interest and speculation, Phobos and Deimos continue to be targets of international inquiry. Coincidental suggestions of their existence can be traced back to earlier popular literature. In his 1727 book *Gulliver's Travels,* Jonathan Swift describes a fictional Voyage to Laputa in which Laputa's astronomers discover two satellites orbiting Mars at distances of 3 and 5 Martian diameters. That vision isn't far off: actual distances of Phobos and Deimos are 1.4 and 3.5 Martian diameters, respectively.

Possibly influenced by Swift, Voltaire's 1752 short story *Micromegas* also refers to two moons of Mars being revealed by an alien visitor to Earth. Accordingly, craters on Deimos are named Swift and Voltaire in recognition of each writer.

In 2011, the Russian NPO Lavochkin Aerospace Company and Space Research Institute launched a Phobos-Grunt (literally Phobos-Ground) sample return mission which failed due to a rocket malfunction prior to leaving Earth orbit. The rocket also carried a 250-pound (113 kg) Chinese Mars orbiter Yinghuo-1, which was to remain in Martian orbit for a year to study the external environment and a Living Interplanetary Flight Experiment to test whether selected organisms can survive for a few years in deep space.[204]

Phobos-Grunt was planned to collect and return about 7 ounces of soil from Phobos. A robotic arm with a pipe-shaped tool split to form a claw would collect samples up to 0.5 inches (1.27 cm) in diameter. A piston on the tool would push the samples into a cylindrical container to then be moved inside a special pipeline into the return module by inflating an elastic bag within the pipe. This process would perform 15 to 20 scoops over a period of 2-7 days.[205]

Since soil characteristics were uncertain, the Phobos lander also provided a drilling extraction device in case the soil was too rocky for scooping. Non-return experiments would continue in situ for a year, using soil heating and sensor equipment to analyze emission spectra.

A sample return stage mounted on top of the lander would accelerate to 22 mph (10 m/s) using springs to escape Phobos' gravity at a safe height before igniting engine for transfer to Earth. A conical-shaped descent vehicle was provided for a hard landing in Kazakhstan without a parachute.

Part Seven: Buzz Words, Charting Pathways Forward

MANY DECADES HAVE passed since I climbed out of the cockpit of a supersonic F-100 armed with nuclear weapons, became an MIT egghead, and then a space traveler. Nowadays, my dedication, indeed my passion, is focused on forging America's future in space, guided by two principles: continuously expanding human presence in space; and global leadership in space.[206]

What should we be reaching for now…and why? Space leadership, technology development, private-public teaming, free-market savvy, and national security prominence…those attributes still define us, or should define us, as a nation.

Apollo 11 symbolized the ability of this nation to conceive a truly pathbreaking idea, prioritize it, create the technology to advance the idea, and then ride it to completion. Apollo was built on the proficiency and professionalism of thousands of dedicated Americans. It was also built on faith and a national commitment.[207]

Apollo is a case where we got it right. If we are to resurrect the profound feeling of participation that accompanied Apollo, we will need a Kennedy-like commitment to human exploration.[208]

So, what is—or should be—the next goal for the American space program?

From a scientific, technology-advancing, meaningful, and politically inspiring point of view, in my opinion, it should be Mars.[209] In reaching outward with method and intent to Mars, and helping others go where we have already gone, America is once again in the business of a momentous and future-focused space exploration program.

We can dare to dream again and to lead. Let us challenge NASA, challenge the White House to think bigger, challenge ourselves to look beyond the moment, and inspire again an entire nation in a way that is evocative, at a time when our country is ready for real inspiration, challenge, leadership, and achievement.

Let's roll…and roll up our sleeves and begin.

University of Houston, Sasakawa International Center for Space Architecture
Mars Base Concept by Canaan Martin

1. Buzz Aldrin. Source: NASA

2. Buzz with graduate students at the University of Houston, Sasakawa International Center for Space Architecture (circa-1980). Source: SICSA

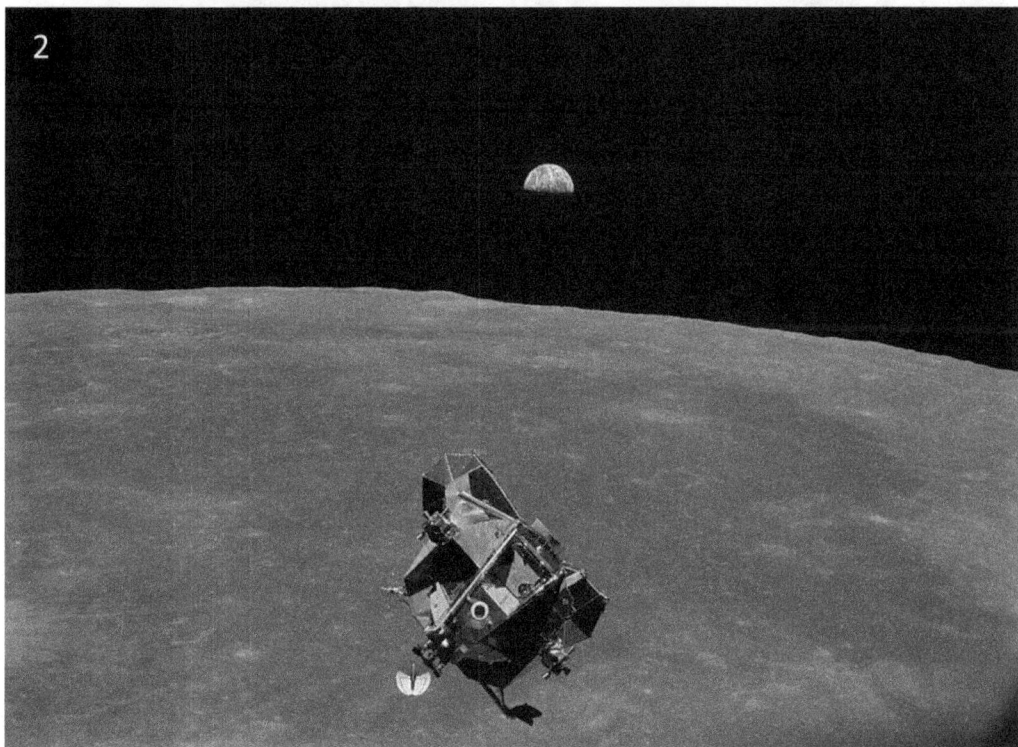

1. Apollo Command Module
   Source: NASA

2. Lunar Ascent Stage, and Lunar
   Descent stage
   Source: NASA

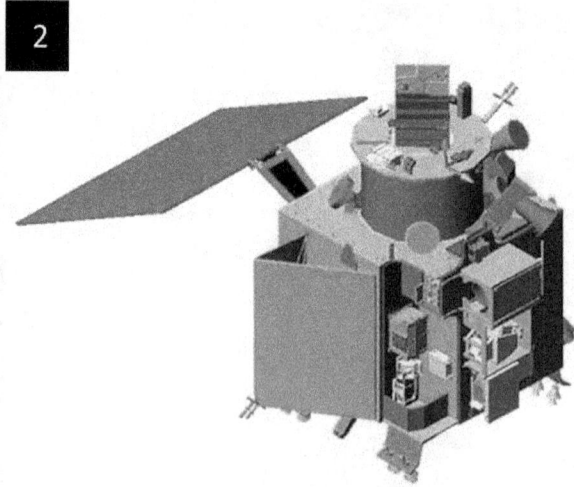

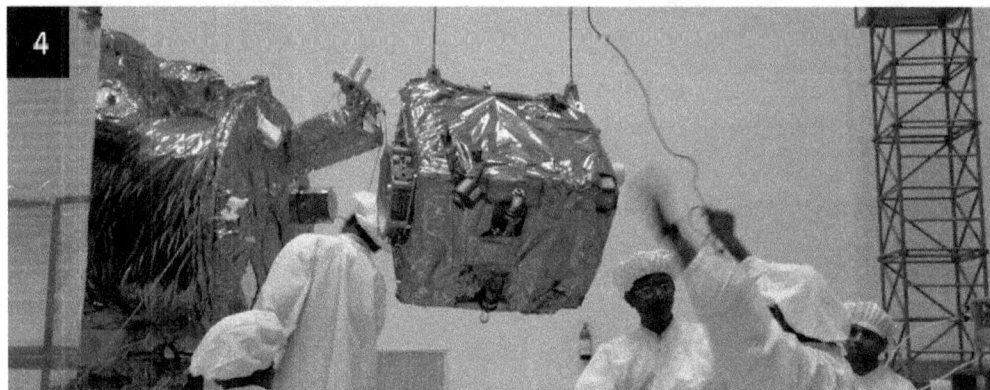

1. NASA aerospace engineer John Houbolt, principal advocate of the Lunar Orbit Rendezvous technique. Source: NASA
2. Chandrayaan 1 Source: Vikram Sarabhai Space Center.
3. Moon minerology mapper Source: NASA
4. Moon impact probe Source: Indian Space Research Organization.

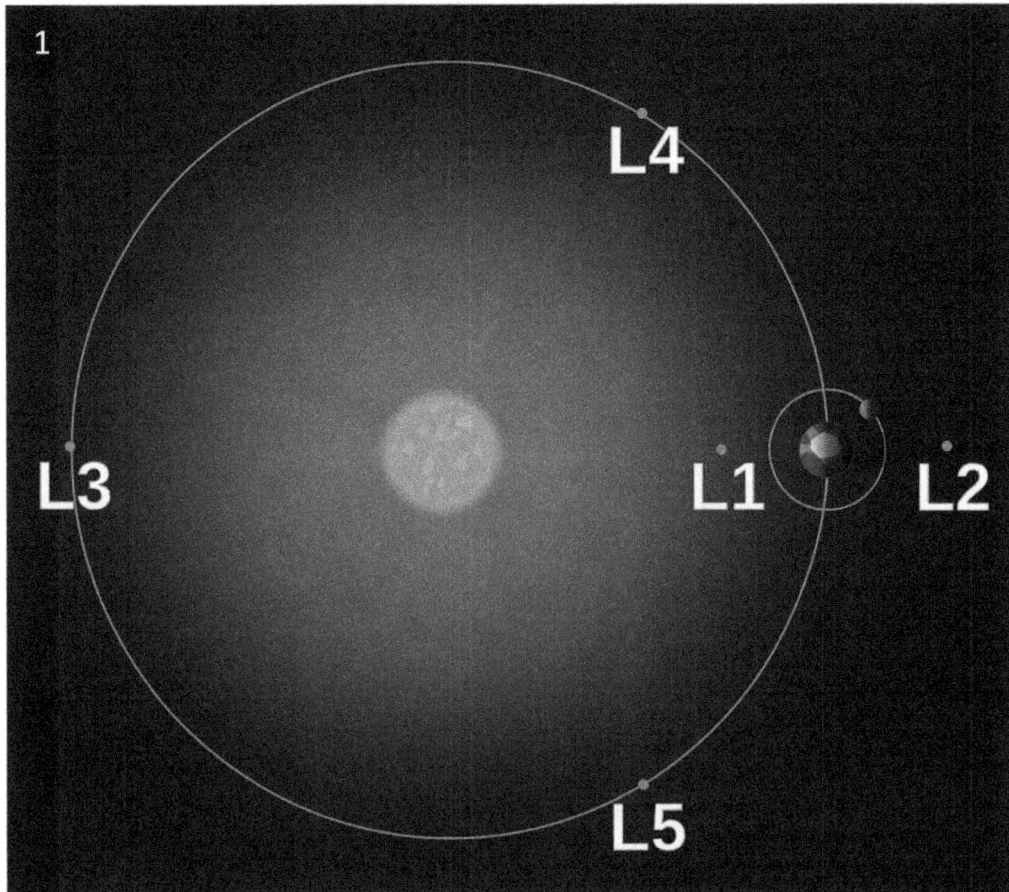

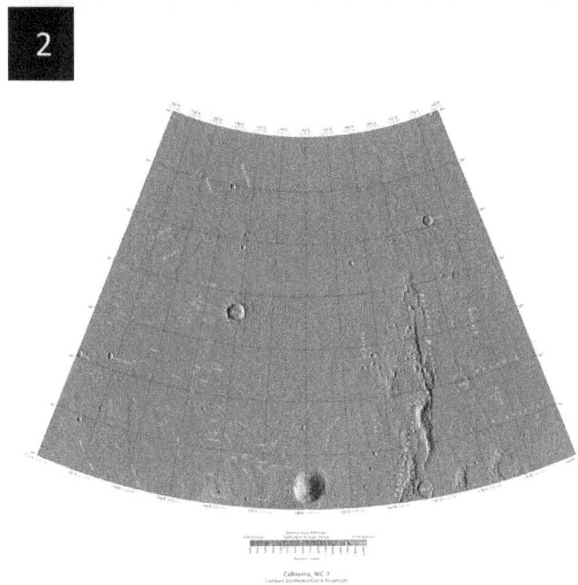

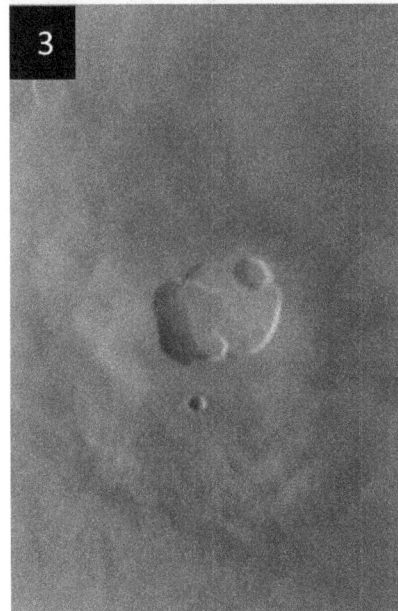

1. Lagrangian Points
   Source: Xander89

2. Cabrenia Quadrangle
   Source: USGS

3. Olympus Mons
   Source: NASA

Chapter Twenty-Two: The Sky is No Longer a Limit

I'VE ALWAYS BEEN conscious of the fact that I was simply the right guy in the right place at the right time. I've always felt fortunate to have had the tremendous opportunity afforded me—lucky to have been born an American, privileged to have been selected as an astronaut, and very blessed to have been a member of the first crew to land on another celestial body.

Some close friends at that time were far less fortunate. In 1966 while I was involved in the Gemini program, my good flying pal Sam Johnson was shot down during his 25th combat mission and incarcerated in a desolate North Vietnamese prisoner of war camp under inhumane living and treatment conditions.

I wore a metal bracelet with Sam's name, rank and serial number on it, along with the date he was shot down, on my trip to the Moon and back. Sam didn't learn about his former wingman buddy being aboard Apollo 11 until his release after seven years in that Hellhole, including 42 months of solitary confinement.

Sam, who passed away in 2020, proudly represented his Dallas, Texas district in the U.S. House of Representatives until 2019.

Another friend I first met flying in that same Korea F-86 group is Mercury astronaut John Glenn. John shot down three Russian MiGs near the Yalu River, one more than I did. He later commented regarding his initial space flight:

> As I hurdled through space, one thought kept crossing my mind; every part of this rocket was supplied by the lowest bidder.

I am deeply honored to have traveled, along with two wonderfully capable Apollo 11 companions, where no humans had previously gone before. Others who made that possible and who build upon that marvelous achievement will journey much farther—as far as human vision, ingenuity, curiosity, and determination will transport them.

So, I'll begin this part of the story with some personal history about how it came about that I so fortunately serve as living proof that the sky is no boundary to human voyages of dreams and destinations.

Born in 1930 and raised in Montclair, New Jersey, I finished high school there. Aviation was pretty much in the family. When I was all of two years of age, my dad took me on my first flight, the two of us winging our way from Newark down to Miami to visit relatives.

My aunt, in fact, was a stewardess for Eastern Airlines. The Lockheed Vega single-engine plane that I flew in was trimmed in red paint to look like an eagle. How could I have grasped then as a child that decades later I would find myself strapped inside a very different breed of flying machine—Apollo 11's lander, the Eagle, en route to the moon's Sea of Tranquility? [210]

My father was both a great friend and inspiration to me, a world-renowned pilot who encouraged me to pursue my own dreams of flying. And while I never really considered this passion to be particularly risky, it has led to much adventure in my life that would definitely fall into that category.

Thanks to wonderful preparation, I have not only survived all of those adventures, but deeply value the life enriching lessons and friendships those experiences have afforded me as well.

Dad had gone to Clark University in Worcester, Massachusetts. His physics professor was Robert Goddard, regarded as the father of liquid-fueled rocketry.[211] My father also attended MIT where he wrote his doctoral thesis on the subject of spinning airplanes.

He loved to fly and spent 38 years in the Army Air Corps (later renamed the U.S. Air Force). He was an outstanding aviator instructor during World War I and was a friend of another MIT graduate, Jimmy Doolittle who led the first carrier-based bomber attack on Tokyo on April 18, 1942 off the USS Hornet.

Dad later helped pave the way with introductions leading to Jimmy's acceptance as an MIT doctoral candidate. Although dad had urged me to attend the Naval Academy, some friends influenced me to choose West Point. That was in 1947 when I was 17 years old.

I later joined the U.S. Air Force where I earned my wings as a pilot of the swept-wing F-86 Saber jet. During the Korean War, Life Magazine featured a picture of the first Soviet-made MiG-15 I destroyed showing the pilot having ejected.

After graduation from high school, I took the West Point motto to heart, Duty, Honor, Country. It's a maxim that remains part of me today.

Surrounded by the influence of aviation, I entered the U.S. Air Force after graduating from the Military Academy. After fighter pilot training I was stationed in Korea, where I flew 66 combat missions in my F-86 Sabre, shooting down two enemy MiG-15 aircraft.[212]

Some of them chased me too, and one very nearly succeeded in ending that activity when my gun jammed. A second MiG encounter north of the Yalu River resulted in a violent series of low altitude turn reversals where I managed to gain the advantage. My gun camera film jammed after scoring hits with my six .50 caliber weapons, but I watched as a second lucky pilot bailed out before his aircraft hit the ground.

Several other NASA astronauts were seasoned Korean War fighter pilots as well. My Apollo 11 partner and friend Neil Armstrong flew 78 combat missions, Wally Schirra flew 90, Gus Grissom flew 100, and Jim McDivitt flew 145. I flew 66 as the war ended.

In the late 1950s the Cold War was escalating between the then Soviet Union and the United States. To be sure, tensions were high. While posted in Germany, I learned of the Soviet's surprising technological feat—the launch of Earth's first artificial satellite in October 1957, a 184-pound (83.46 kg) sphere called Sputnik.

As the import of Sputnik sank in, against the backdrop of the Cold War, the political and public reaction spurred on the space age. It became the starting gun for the space race, leading to the creation of NASA the following year.[213]

The Soviet Union achieved yet another triumph on April 12, 1961, by sending the first human into Earth orbit, cosmonaut Yuri Gagarin, in his Vostok 1 spacecraft. As a comparative note, a few weeks after Gagarin's mission of 108 minutes duration, NASA flew on May 5 America's first Mercury astronaut, Alan Shepard, on a 15-minute suborbital flight that touched the edge of space.

A mere 20 days after Shepard's mission, President John F. Kennedy boldly challenged America to commit itself to achieving the goal of landing a man on the moon before the end of that decade.

Many of those at the helm of a newly formed NASA thought the challenge to be impossible. The know-how just wasn't there. The nation had little more than 15 minutes of spaceflight experience under its belt.

But what America did have was the President with vision, determination, and the confidence that such a goal was attainable. By publicly stating our goal and by establishing an explicit time period on a very clear accomplishment, President Kennedy offered no back door. We either had to do it or not make the grade… and no one was interested in failing. Even then, failure was not an option.[214]

If space was going to be our next new frontier, then I wanted to be a part of getting there. After completing my tour of duty in Germany, I decided to continue my education and receive my PhD in astronautics from the Massachusetts Institute of Technology, the same university my father had gone to.

For my thesis, *Guidance for Manned Orbital Rendezvous*, I adapted my experience as a fighter pilot intercepting enemy aircraft to develop a technique for two piloted spacecraft to meet in space. I dedicated that final paper to the American astronauts.[215]

The first time I filled out the forms to be a NASA astronaut, my application was turned down. I was not a test pilot.

Determined to seek a career as an astronaut, I applied again. This time, my jet fighter experience and NASA's interest in my concept for space rendezvous influenced them to accept me in the third group of astronauts in October 1963. I became known to my astronaut peers as Dr. Rendezvous.[216]

It's essential to note the insertion of the Gemini program. It was a fundamental stepping-stone, a bridge between the one-man Mercury and three-person Apollo programs, primarily to test equipment, to do trial runs of rendezvous and docking scenarios in Earth orbit, and to train astronauts and ground crews for future Apollo missions.

U.S. Air Force training, experience and mental focus on the challenge at hand greatly benefited me as a prelude to Gemini and Apollo. Yes, and lots of good luck helped as well. My graduate PhD thesis topic came in particularly handy during the 1966 Gemini 12 mission.

As Jim Lovell piloted the fighter aircraft-like spacecraft, I executed three successful spacewalks for a record total of 5½ hours traveling at 17,500 miles per hour above Earth in my own tiny one-man pressure suit.

My MIT orbital rendezvous studies were also extremely valuable in preparation for my Apollo 11 trip to the Moon and back. Neil and I didn't want to botch the Eagle Lander rendezvous with Mike Collins in the Columbia spacecraft that took us home.

On November 11, 1966, I made my first spaceflight as pilot of Gemini 12, alongside James Lovell, the mission command pilot. That nearly four-day flight brought the Gemini Program of ten piloted missions to a successful close.

During the flight, I was able to establish that new record for spacewalking. To be honest, up to that point, we had failed miserably in the Gemini program to show that an astronaut could easily and effectively work outside a space vehicle.

We used microgravity training in parabolic flights of airplanes, but that didn't solve the Gemini spacewalking problems at all. It took underwater training that I introduced which became a standard fixture in simulating extravehicular activity (EVA) here on Earth. Thanks to that underwater training and the use of appropriate restraints, I chalked up my successful EVA without taxing my space suit.[217]

I just couldn't wait to get into my personal preference kit and get my small slide rule out and have it float there in front of me. Being a pipe smoker at the time, I also brought my pipe along and put it in my mouth (unlighted, of course), with Lovell taking a picture of that episode.[218]

Yes, Gemini was the link that prepared us for the Apollo missions to the Moon, but we still had major work to do.

In all, there was a team of 400,000 people enabling a common goal…those NASA managers, engineers, and technicians working side-by-side with industry contractors who designed and built the multistage Saturn V booster needed to propel us to the Moon.[219]

It was a unified enterprise, a synergy of innovation, effort, and teamwork that was unstoppable to transform a long-held dream into a reality.

In responding to President Kennedy's bold challenge of landing a man on the Moon by decade's end, there were many alternatives discussed as to how we could get there and safely return to Earth.

A very gifted NASA engineer, John Houbolt, trumped even the revered U.S. space program leader, Wernher von Braun, who favored a huge monstrous rocket, a multipurpose spacecraft, and direct flight to get to the Moon and back.[220]

Houbolt backed a lunar-orbit rendezvous plan. It called for not a multipurpose crew vehicle architecture, but instead, a segmented way to achieve the Moon landing feat. When the Apollo Moon landing method was finally scripted, it adopted segmentation of the mission: using an Apollo Command Module as discreet from the Service Module and separating the Lunar Ascent Stage from the Lunar Descent Stage.[221]

Houbolt's master plan became a plus for me in terms of my MIT rendezvous work. The critical key to this approach would be our ability to reliably rendezvous two spacecraft in orbit around the Moon, a very dangerous maneuver. For if that rendezvous failed, there would be no way to rescue the astronauts. Luckily, that MIT expertise was exactly what was required.[222]

Eight years after President Kennedy committed us to strive for the impossible, Neil Armstrong and I walked across the sun-drenched terrain of the Moon.

Nearly a billion people all over the world watched and listened as we ventured across that magnificent desolation. With Mike Collins circling above us, and even though we were farther away from our planet than any three humans had ever been, we felt connected to home.[223]

Yes, Apollo 11 was historic, but it was fraught with risks. When we finally set the Eagle lander down, with Neil piloting and me calling out descent numbers for him, we had only an estimated 16 seconds of fuel left in the Descent Stage.

On the surface, if we had fallen and torn a suit, there wasn't much chance of survival. If the one ascent engine didn't ignite—or if the onboard computer had a glitch—we would never have left the Moon. If the rendezvous with Mike Collins circling the Moon in the Command Module hadn't gone flawlessly, we then would have faced rather nasty consequences.

That's just a few of a string of "ifs."[224]

Just in case any of those ifs occurred, a July 18, 1969 internal White House statement titled *In Event of Moon Disaster* included this ominous phrasing:

> *Fate has ordained that the men who went to the Moon to explore in peace will stay on the Moon to rest in peace.*[225]

By the way, while Neil was the first human to step onto the Moon, I'm the first alien from another world to enter a spacecraft that was going to Earth.[226]

Chapter Twenty-Three: What Human Exploration and Development Does for America

FIRST OF ALL, human space exploration and development reminds the American public that nothing is impossible if free people work together to accomplish great things.

It captures the imagination of our youth and inspires them to study science, technology, mathematics, and engineering.

Furthermore, a vigorous human spaceflight fuels the American workforce with high technology and cutting-edge aerospace jobs. And it fosters collaborative international relationships to ensure U.S. foreign policy leadership.[227]

An essential component of the future is to maintain U.S. leadership of human space transportation. America must lead where it matters most, providing the systems to safely transport people across space.

We cannot afford to toss away over 50 years of accumulated experience and knowledge of human spaceflight systems. Once lost, it will require decades to replace. Moreover, without stated, clear destinations or goals for human spaceflight beyond Earth orbit, what is created is major uncertainty over the future of the United States in space.[228]

Our Nation's Youth

Space exploration and development inspires and energizes education of America's youth in the areas of science, technology, engineering, and math (STEM) that is critical to our future.[229]

Some of those taking part will become the scientists, engineers, and industrialists who build the rockets and spacecraft of the future; some will develop into professional astronauts and colonizers of other worlds; and many will turn into citizen explorers—young global space travelers—who venture into Earth orbit, holiday at space hotels, and, not too far in the future, take even longer journeys.[230]

There are several reasons space is a powerful tool to make STEM education more global, equitable, affordable, creative, attractive, and adaptable. These students will learn that space is inherently borderless, belongs to everyone, and is a fast-growing and promising industry.

Additionally, space-related STEM content appeals to students of all ages; inspires and motivates creativity; develops curiosity and critical thinking; is interdisciplinary; appeals to both genders and promotes equality; promotes international and cross-cultural cooperation; and strives for a common, thriving future.[231]

Let's also add art creativity and perspective to that mix—making it science, technology, engineering, *art*, and mathematics—STEAM power for short.[232]

The space voyages beyond our Earth over the next 25 years will also motivate the next wave of technology entrepreneurs. This search for new horizons will enhance America's global leadership and encourage international cooperation among spacefaring nations.

Humanity is destined to explore, settle, and expand outward into the universe. Homesteading our Solar System is a reach outward to what lies beyond—other planets.

But doing so urgently requires a rekindling of America's space program to ignite a new wave of support and participation in the United States and elsewhere.

This is a spot-on space trek of inspiration and aspiration for future generations.

Younger readers have probably heard their parents or grandparents say: "The world is yours." I want to take it one step further and say: "The worlds are yours." [233]

When I was a young person, I certainly wasn't the only one in the neighborhood who looked upward and dreamed about going to the Moon or stepping onto other planets. I was a reader of science fiction.

At that moment in time, no one had traveled into space. Everyone, including me, had to bank on imagination to conjure up ways to make those dreams come true.[234]

To the younger reader, will you be one of the first people to walk on Mars?

You could even be among the first human settlers to colonize that planet. There are out-of-this-world things to accomplish—all fostered by the ability to reach for places that no one has ever reached for in the past.[235]

International Cooperation

Space exploration and development also offers opportunities for peaceful and productive international collaboration. Despite the Cold War tensions between the United States and the former Soviet Union that characterized the space race of the 1960s, the Russians have become critical partners in the International Space Station—a collective effort of 16 nations.

It is time now to inspire the international community to jointly explore and develop the Moon as a partnership. Forget the space race. That is now a mode that's outmoded. We can afford to be magnanimous. America was first to set foot on the Moon. Now let us make it a first step for all humankind.[236]

How do we frame our collaborative or international effort to get to the Moon again? Let me reemphasize: Certainly not as a competition. We have done that, and to restart that engine is to return to a race we won. Let's take a pass on that one. Do not put NASA astronauts on the Moon. They have other places to go.[237]

America should chart a course of being the national leader of this international activity to develop the Moon, but not by spending money placing U.S. government people on its surface.

Our focus should be limited to robots on the lunar surface that are dutifully employed to do scientific, commercial, and other private-sector work. We need to provide the non-surface lunar infrastructure and make that available to other governments—China, India, and others.

Lunar-derived surface water, for example, can support human activities and also provide a source of fuel for Mars-bound rockets. Refueling landers in Low-Earth Orbit (LEO) can provide great economic efficiency benefits over launching that fuel from the Earth's surface because of the Moon's lack of atmospheric friction and much weaker gravitational resistance to overcome.

Apollo 17 astronaut Harrison Schmitt—the last human to step on the Moon's surface—is an advocate for Helium-3. Perhaps one day if and when atomic fusion proves to be a viable energy source, He-3 will fuel electricity for bases and operations on the Moon and Mars.

Chapter Twenty-Four: Applying Lessons from Antarctica

IN MANY WAYS, I anticipate an outpost on the Moon that could mirror what has already taken place here on Earth, in Antarctica.

Americans have been studying Antarctica and its interactions with the rest of our globe since 1956. Visiting researchers delve into glaciology, biology and medicine, geology and geophysics, oceanography, climate studies, astronomy, and astrophysics.

Contractors and units of the military provide operational support at year-round stations: Palmer Station, Amundsen-Scott South Pole Station, and McMurdo Station (the main U.S. station in Antarctica). Today, the U.S. Antarctic Program supports a peak population of about 1,600 men and women.[238]

The tab for the U.S. Antarctic Program—an effort that supports scientific research in Antarctica and the waters surrounding, with the goals of fostering cooperative research with other nations, protecting the Antarctic environment, and conserving living resources—is picked up by the National Science Foundation.

On my own trek to Antarctica, I could clearly see an analogy between five-plus decades of research in that icy wasteland and what awaits us on the Moon. There's a long line of people who hunger to travel to Antarctica and carry out research tasks. Seeking answers to questions leads to new inquiries.

The Moon is just as complex, as remarkable, and as fruitful in an exploratory sense as Antarctica.[239]

Over the last several years, for instance, a number of different lines of international evidence have pooled together to help shore up the case for water on the Moon.

Evidence of lunar water was first discovered by NASA's 1971 Apollo 14 mission.

In 2008, an Indian Space Research Organization's Chandrayaan-1 spacecraft carried a NASA instrument, the Moon Mineralogy Mapper, that found evidence of water molecules on the lunar surface.

The Chandrayaan-1's Moon Impact Probe (MIP) also sensed that it flew through an exospheric "water cloud" during its plunge onto the lunar landscape. What the MIP found might have been water actually in motion that migrates and concentrates in the ultracold, permanently shadowed lunar craters.

Then there were the NASA Lunar Crater Observation and Sensing Satellite (LCROSS) observations in 2009 that detected water vapor and ice particles kicked up after the LCROSS Centaur upper stage was purposely slam-dunked into the Moon.[240] NASA's SOFIA observatory again found evidence of lunar water in 2020.

And here is another appealing link between Antarctica and the Moon. Shackleton crater is a large and deep impact feature that lies at the south pole of the Moon. This crater was named after Ernest Henry Shackleton, the intrepid Anglo-Irish explorer who took part in the period later labeled as the Heroic Age of Antarctic Exploration, a time span that stretched from the end of the 19th century to the early 1920s.

The shadowed portion of the crater was scanned with the Terrain Camera on board the Japanese SELENE spacecraft in October 2008, helping to gauge the slopes and central peak of Shackleton crater. Those observations were followed by the launch in 2009 of NASA's Lunar Reconnaissance Orbiter. It has played a significant role in eyeing Shackleton with radar and an array of other sensors.[241]

Shackleton crater is more than 12 miles wide and 2 miles deep, about as deep as Earth's oceans. The peaks along the crater's rim are exposed to almost continual sunlight, while its interior is forever in shadow.

All this adds up to this captivating feature being an ideal spot for the International Lunar Research Base situated on the edge of the crater. Shackleton hosts both regions of near-permanent darkness and near-permanent sunlight, just the thing for Sun-energized power stations. And like real estate here on Earth, it's all about location.[242]

Having water ice within sun-shy Shackleton raises the outlook of harvesting those cold-trapped deposits, an extraterrestrial commodity that would minimize the need to carry water from Earth to the Moon. Not only can it be processed for human consumption, it can also be transformed into fuel.

Yet another bonus about this crater is that roughly 72 miles away is Malapert Mountain, a peak that is perpetually visible from Earth and can be topped by a radio relay station.[243]

While there is mounting consensus regarding Shackleton as a future encampment, resolution of the ice issue is likely to require more on-the-spot survey work by robotic craft.[244]

The Moon is close, it is interesting, and it's useful. As the rocket flies, traversing cislunar space—traveling from Earth to the Moon—takes just three days. Additionally, the Moon contains a record of planetary history, evolution, and processes unavailable for study on Earth or elsewhere.

In terms of its usefulness, projects at the Moon can help to retire risk for future planetary missions—say sending people to Mars or to the asteroids—by sharpening our space skills and putting to the test exploration hardware for future deep space sojourns.[245]

In any case, go-it-alone initiatives create the prospect of duplication of effort—and the wasteful use of resources. It is time to build on each nation's talents and reduce mission risk by sharing information and capabilities.[246] The best plan is to cooperate with international partners who also want to reach the Moon, to offer a hand. Our resources must be saved and spent on moving toward establishing human permanence on Mars.[247]

Chapter Twenty-Five: The Case for Mars

MARS, THE RED PLANET, has long drawn our curiosity—and there's now a rover on its surface named just that.

We first made eye contact with that new world of discovery thanks to Earth-based telescopes. Mars is an intellectual magnet provoking thought. Consider the view of astronomer Percival Lowell, writing in his 1908 book, *Mars as the Abode of Life*:[248]

> *Thus, not only do the observations we have scanned lead us to the conclusion that Mars at this moment is inhabited, but they land us at the further one that these denizens are of an order whose acquaintance was worth the making. Whether we ever shall come to converse with them in any more instant way is a question upon which science at present has no data to decide.*

But science about Mars has proceeded ever since, and since 1960, telescopic-driven talk about life on Mars has been augmented by voyages of numbers of automated spacecraft—sent there by multiple nations.

Mars has been flown by, orbited, smacked into, radar examined, and rocketed onto, as well as bounced upon, rolled over, shoveled, drilled into, baked, and even laser blasted. Still to come: Mars being stepped on.

As always, front and center is the power of Mars to entice us to brood over some key, compelling questions, particularly if life ever was sparked into being there. If so, did it perish or is it still resident on the planet?

Understanding the Martian climate and atmosphere, including the evolution of Mars's surface and interior, can be looped back into grasping the past, present, and future of Earth.

The geologic record of early Mars has been preserved, chronicling the period more than 3.5 billion years ago when life is likely to have started on Earth—a time period whose record is mostly missing on our own planet. In addition, Mars exploration can allow us to turn back the clock and see if life arose elsewhere in our solar system neighborhood.[249]

But why send humans to Mars in the first place? There is common agreement that humans trump machines in many ways. They offer speed and efficiency to perform tasks. On-the-spot astronauts offer nimbleness and dexterity to go places that are challenging for robots to access. Then there are the innate smarts, ingenuity, and adaptability of a human to evaluate in real time a situation, then improvise to prevail over surprises.

Now, and in the near future, robotic exploration of Mars is providing a window on a world that can be a true home away from home for future colonists. The first footfalls on Mars will mark a historic milestone, an enterprise that requires human tenacity matched with technology to anchor ourselves on another world.

Exploring Mars is a far different venture from Apollo expeditions to the Moon; it necessitates leaving our home planet on lengthy missions with a constrained return capability.

Once humans are at distant Mars, there is a very narrow window within which it's feasible to return to Earth—a fundamental distinction between our reaching Earth's Moon in the 1960s and stretching outward to Mars in the decades to come.[250]

All this is preface to a major judgment—one that I feel NASA planners are dodging. There is no reason to make a humans-to-Mars program look like an Apollo Moon project.

We need to start thinking about building permanence on the red planet, and what it takes to do that. I feel very strongly about this.

This is an entirely different mission than just putting people on the surface of that planet, claiming success, having them set up some experiments and plant a flag, to be followed by quickly bringing the crew back to Earth, as was done in the Apollo program.[251]

What are you going to do with astronauts who first reach the surface of Mars and then turn around and rocket back homeward? What are they going to do, write their memoirs? Would they go again? Having them repeat the voyage, in my view, is dim-witted. Why don't they stay there on Mars?

Earth isn't the only world for us anymore.

Elon Musk has said that he wants SpaceX to reach Mars so that humanity is not a single-planet species.

I suggest that going to Mars means permanence on the planet—a mission by which we are building up a confidence level to become that two-planet species.[252]

Establishing a footing on distant Mars is a complex operation.

The challenge ahead is monumental and historic. We are on a pathway to homestead the red planet courtesy of robotic explorers that are surveying what now looks like unreal real estate.

Other than our limited trips to the Moon via Apollo, humans have never embarked upon a mission that's on a par with marching off to Mars. The best analogs so far are Antarctic, undersea, and International Space Station expeditions, but these are distant cousins to the isolation, remoteness, and challenges that will be faced by courageous men and women stationed on Mars, many millions of miles from Earth.[253]

A NASA Mars reference document emphasizes the need for more study of the composition of a Mars crew, based on personal and interpersonal characteristics "that promote smooth-functioning and productive groups, as well as on the skill mix that is needed to sustain complex operations."[254]

Here, there are many behavior, performance, and human factor unknowns. Living far from Earth in a remote and confined environment will surely induce physiological and psychological stresses which are sure to haunt the first humans on Mars—including isolation and loss of privacy.[255]

Nonetheless, there's also a certain familiarity with remote Mars.

It did not go unnoticed that the first color images transmitted from the Curiosity rover showed layered buttes and other features reminiscent of the southwestern United States. There is an evolving comfort level with Mars. It is a perspective that beckons us to push forward.[256]

Chapter Twenty-Six: Key Technical Priorities and Challenges

MARS EXPLORATION AND development will require solutions to virtually endless human and equipment challenges.

Here's a quick look at what I view as the Buzz Basics—a list of the necessary technological developments required for moving outward and onward.

Space radiation protection will be required to safeguard astronauts from solar particle events, galactic cosmic rays, and radiation trapped in planetary magnetic belts or encountered on a planetary body's surface.

Long-duration missions present an urgent need to tackle this issue, perhaps by using electrostatic or magnetic force radiation shielding, use of new lightweight materials, or adoption of antiradiation pharmaceuticals to thwart, alleviate, or restore to health any damage suffered by the crew exposure.[257]

Life support for crews on long-haul space travel also mandates the need for reliable, closed-loop environmental control and life-support systems.

We must learn how to maximize self-sufficiency and minimize the need for resupply of vital consumables—air, water, and food. As crews move distant from our home planet, the current approach of regularly resupplying life-support consumables and returning wastes to Earth will not be possible.

Redundant systems are crucial in the event of failure, whether it's for a mechanical or software problem.

Above all, human-critical applications, such as flight control and life support, must not fail as we venture outward on long-duration missions beyond Earth. Reliability through redundancy and backup must be a priority, but also more attention should be paid to systems that can be readily fixed—if properly designed to be fixable.

Portable, deployable pressurized inflatable structures which are currently being developed offer means to provide more roomy habitats to improve crew comfort, morale and operations on extended space voyages and on lunar and Mars surfaces.

Landing systems are a critical technology, be it for Earth reentry and/or robotic or human Moon or Mars landings.

We must amplify the ability to land at a variety of planetary locales and a variety of times. Precision landing capability allows a spacecraft to land closer to a specific, predetermined position for safety's sake, whether on autopilot or crew control, as well as maximizing operational or science objectives. In other words, the closer you are to where you want to be, the better.[258]

Here, aerobraking is a technique used to reduce velocity of a spacecraft into orbit around a planet or its moon by using the atmosphere, where friction causes the spacecraft to slow down. This permits a quick orbital capture of a spacecraft at Mars, reducing the need for hauling a load of onboard propellant.

Aerobraking technologies, however, support only the entry phase. There's also a need for on-the-spot delivery of heavier and heavier payloads on bodies of interest, particularly in the case for human exploration of Mars.

On one hand, humans bring perception, speed and mobility, dexterity, and an inquisitive nature. On the other hand, robots are able to cope with the surly climes of Mars while carrying out boring and risky jobs.

The first Mars base facilities will be built before human occupancy using remotely operated telerobotic systems. Human-controlled robotic equipment will also be necessary to extract important resources such as the water for human consumption and for hydrogen-oxygen rocket fuel.

In addition, remote telepresence making use of low-latency communication designed to reduce transmission delays over links will afford human cognition the appearance of being there to scout out mining opportunities, and pre-position habitats without need of on-site, space-suited astronauts.

Chapter Twenty-Seven: Strategic Pathways Forward

WHEN NEIL AND I stepped upon the surface of the Moon at Tranquility Base, we fulfilled a dream held by humankind for centuries. As inscribed on the plaque affixed to the ladder of our lander: "We Came in Peace for All Mankind."

It was, truly, one small step. But more steps are needed. There is no compelling reason to forgo our longer-term goal of permanent human presence on Mars. Consequently, great care must be taken that precious dollar resources needed for the great leap to Mars are not sidetracked to the Moon.[259]

The United States has more experience at the Moon than any other nation. The country made a huge expenditure in the 1960s and 1970s to gain that leadership.

So, to just toss that investment away, Is ridiculous.

However, what we now need to do is foster a presence at strategic locations in space called Lagrangian points that permit the United States to robotically assemble, piece by piece, hardware and habitation on the Moon.

There are five Lagrangian points in the Earth-Moon system—as well as in the sun-Earth system—in space where the combined gravitational forces and the orbital motion of two bodies balance each other. This condition enables spacecraft in these locations abilities to linger without consuming much precious rocket fuel.[260]

The Earth-Moon Lagrangian points known as E-M 1 (L-1) and E-M 2 (L-2) are viable locations for lunar communications, places where the combined gravity of Earth and the Moon permits a spacecraft to be synchronized with the Moon in its orbit around Earth.

In other words, the spacecraft appears to hover over the far side of the Moon, affording crew members continuous line-of-sight visibility to the entire far side of both the Moon and Earth.[261]

A pair of communication satellites in the halo orbits circling around the Earth-Moon Lagrangian points 1 and 2 will also provide radio blackout-free coverage of spacecraft in lunar orbit and for most of the lunar surface.[262]

This communication relay system can tackle the challenge of contact with the lunar far side, which is blocked from direct line of sight with Earth.

Matching Earth-Moon Lagrangian points with astronauts operating telerobotic hardware allows the assembly of infrastructure on the Moon, carrying out surface science, scouting out and unearthing important lunar resources.

A lunar navigation system consisting of a constellation of perhaps four or five satellites can provide the precise navigation needed to make lunar research much more effective and less risky, both for teleoperated rovers and for human explorers.

I've been there. Working on the Moon is not easy.

You're faced with a lack of reference points and landmarks. The Moon is such a small body, the nearness of the lunar horizon makes navigation on the lunar surface tricky. It's very easy to get lost on the surface of the Moon, particularly if you are in rough terrain—the very type of landscape that is likely to be most attractive for study.[263]

Utilizing the Moons of Mars

At Mars, we've been given a wonderful set of moons, Phobos and Deimos—two different choices—from which we can telerobotically pre-position surface hardware.

Phobos and Deimos are, in a sense, offshore islands of Mars, discovered in 1877 by Asaph Hall at the U.S. Naval Observatory in Washington D.C. They were tagged with names from Greek mythology: Phobos means fear, Deimos, terror. In the future, these Martian moons are likely to symbolize just the opposite: courage and security.

Both moons are tidally locked to Mars, as our own Moon is relative to Earth: Phobos and Deimos present the same side to Mars all the time.

Phobos is the innermost moon of Mars, only 16.7 miles (26.9 km) in diameter but the larger of the two moons. Diminutive Deimos is a little over 7 miles (11 km) in breadth.

Scientifically, both Martian moons are oddballs. There is continual dispute as to where they came from. Just how did they get there?

Conjecture about them being captured asteroids or congregated with Mars is debatable. These two objects are a cosmic detective story, and we need more clues to sort out their true nature.[264]

The good news here is that Phobos orbits Mars at just 5,827 miles (9,377 km) from the planet's surface. It circles Mars in about eight hours. It is nearer to its parent planet than any other known moon in our solar system.

Since Phobos hurtles around Mars faster than the planet rotates, future Mars-walkers could see this moon is bathed in reflected light off of the red planet. This Mars-shine is akin to Earthshine, when sunlight reflects off our planet and illuminates the Moon's night side.[265]

Phobos is a heavily cratered, irregular body with no atmosphere. The gravity field is very weak—less than one-thousandth the gravity on Earth—so that spacecraft would dock with it rather than actually landing on it.

It will be much easier to depart Phobos than Earth's Moon or Mars. The escape velocity from this moon is just 25 miles an hour.

This moon's most eye-catching feature is Stickney, a six-mile-wide crater. When the object that formed this crater hit Phobos, its impact fashioned streak patterns across the moon's surface.

The day and night sides of the moon have been gauged, showing extreme temperature variations; the sunlit side of Phobos is like a pleasant winter day in Chicago, while only a few miles away, on the dark side of the moon, the temperature is more ruthless than a night in Antarctica.[266]

Taking all these factors and others into account, I feel that Phobos may well be the ideal location from which to support a non-human, hands-off Mars program—at least initially.

Phobos can serve as a way station, a perfect perch where crews can run robotic vehicles and pre-position habitation modules and other equipment on Mars more directly, in a much shorter communication delay time than commands sent from faraway Earth.[267]

I'm not alone in valuing the Martian moons as fundamental to opening up Mars to human visitation. Similar in thought was the late Fred Singer, an emeritus professor of environmental science at the University of Virginia. He was also the founding director of the National Weather Bureau's Satellite Service Center back in 1962 and has a long pedigree of building and flying space instruments.[1]

---

[1] Fred Singer's interest in Phobos and Deimos as possible Mars mission staging locations was also shared by the editor—Larry Bell—who first visited my University of Houston office to discuss the subject in 2007. Fred was also enormously well-

Fred had an enduring enthrallment with Phobos and Deimos and yet he remained perplexed as to how and why the Mars moons came to be.

He favored Deimos as the place to establish a human-tended laboratory. Being higher above Mars, it's easier to get to and is nearly in synchronous orbit, a far better situation from which to observe and operate equipment on the planet below.

We both agreed on the plan for teleoperation of Mars machinery from either Phobos or Deimos because the light-speed distance, even coupled with relay satellites circling Mars, is far shorter than possible between Earth-Mars ground control. This enables a fraction-of-second time delay, within human reaction time on the surface rather than a typical 40-minute delay from Earth.[268]

After all, Mars is vast. It's a huge planet with a lot of real estate, some of it very hazardous in terms of crevasses, caves, steep hills, giant canyons, and high mountains. Better to lose a robot or two than have a person face a deadly predicament.[269]

Setting up a lab/control center on one moon of Mars also allows humans to voyage to the other. This sortie by space taxi would be of great value scientifically, enabling a comparative sampling of both moons. Are they made of the same stuff? Do they have a common origin?

As Fred Singer suggested, we simply don't know.[2]

There's a chance of finding signs of life from Mars ejecta captured by Phobos, a prospect less likely for the outermost moon, Deimos. Consequently, Phobos could be the Library of Alexandria of Mars…akin to the ancient Library in Alexandria, Egypt.

Accordingly, this Martian moon could be a treasure trove, rife with knowledge and record keeping that documents all of Martian history.[270]

Deimos has a site near the "arctic circle" that offers the advantage ten months of continuous sunlight during the Martian summer, enabling the use of simple solar power systems. Astronauts would also have direct lines of sight both to Earth and to rovers on the surface of Mars, simplifying communications.[271]

The view of Mars from Deimos would be stunning. For instance, Olympus Mons alone, the great volcanic mountain on the planet, would be roughly three times wider than the full Moon seen from Earth.[272]

Mars Challenges and Opportunities

There are countless human safety and health risks along with living and work challenges that must be addressed in planning for crewed missions and longer stays on Mars and its moons.

An imperative in exploration of Mars and its moons necessitates a shift in mind-set—not taking everything you need by launching it from Earth. Instead, like pioneering in a new land, you need to find, acquire, and utilize all available resources at destination's end.

In space culture, this is referred to as In-Situ Resource Utilization (ISRU) to obtain all useful products that can be tapped—very much including water obtained from Mars subsurface ice, construction materials on its surface and methane in its atmosphere.

Putting in place an effective ISRU system will lessen the need for resupply missions.

Water that is converted to hydrogen and oxygen for rocket fuel will also require solar power systems, and robotic equipment for excavation as well as means for long-term containment until use.

A major scientific attraction in exploring Mars will be to search for extraterrestrial life. In doing so, careful precautions will be necessary to assure that this exploration process doesn't contaminate the Marian environment.

---

informed about climate science, and he introduced me to that interest. Early satellite observations have shown no tell-tale signs of warming in the Earth's lower atmosphere as predicted by virtual climate models.

[2] Fred and I [Larry Bell] collaborated on climate investigations, an activity that led me to publish two books on the subject, *Climate of Corruption: Politics and Power Behind the Global Warming Hoax* (2011) and *Scared Witless: Prophets and Profits of Climate Doom* (2015). The latter of these books was dedicated to honor Fred Singer who encouraged me to write it.

Conversely, it will also be imperative to avoid any remote possibility that soil samples transferred from Mars might support living organisms that could inadvertently reproduce on Earth to damage some aspect of our biosphere. Avoiding both of these possible eventualities is termed Planetary Protection."[273]

In striving for settlement of Mars, new technologies must be mastered. Agriculture under extreme conditions, power generation, radiation protection, and advanced life-support systems are called for. There will also be a need to take into account ionizing radiation and the toxicity of soils, among other items.[274]

Astronauts exploring Mars will likely bring hydroponic growth labs where vegetables can be grown to provide early mission crews and future settlers with added nutrition and variety.[275]

Once a crew lands, they will need effective and reliable shelter to permit outside excursions. The particular crew might, for example, investigate the Martian surface in a shielded and pressurized exploration vehicle, say for weeks at a time, without returning to the habitat. Strolling Mars-walkers will need protection from radiation and dust to safely survey and work on the surface.[276]

In summary, in the words of Mars Society founder Robert Zubrin:[277]

> *Mars is key to humanity's future in space. It is the closest planet that has all the resources needed to support life and technological civilization. Its complexity uniquely demands the skills of human explorers, who will pave the way for human settlers.*

Zubrin, a creative astronautical engineer, is an energetic, effervescent, vocal, and steadfast spokesperson for putting into high gear what he terms the Mars Direct approach a sustained humans-to-Mars plan that he has scripted in his book highly detailed book *The Case for Mars: The Plan to Settle the Red Planet and Why We Must.*

As I previously noted, Zubrin's blueprint for the red planet makes use of the Martian atmosphere to generate rocket fuel, extracting water from the Martian soil, and eventually using the abundant mineral supplies of Mars for construction purposes. In doing so, his plan emphasizes a pragmatic minimalist live-off-the land approach and uses existing launch technology.

The general outline of Mars Direct is straightforward:

- In the first year of implementation, an Earth Return Vehicle (ERV) is launched to Mars, arriving six months later. Upon landing on the surface, a rover is deployed that contains nuclear reactors necessary to generate rocket fuel for the return trip. After 13 months, a fully fueled ERV will be sitting on the surface of Mars.

- During the next launch window, 26 months after the ERV launches, two more craft are sent up: a second ERV and a habitat module—hab for short, which is the astronauts' ship. This time the ERV is sent on a low-power trajectory, designed to arrive at Mars in eight months—so that it can land at the same site as the hab, in the event the first ERV experiences any problems.

- Assuming that the first ERV works as planned, the second ERV is landed at a different site, thus opening up another area of Mars for exploration by the next crew.

- After a year and a half on the Martian surface, the first crew returns to Earth, leaving behind the hab, the rovers associated with it, and any ongoing experiments conducted there. They land on Earth six months later to a hero's welcome, with the next ERV/hab already on course for the red planet.

- With two launches during each launch window—one ERV and one hab—more and more of Mars will be ready for human occupancy. Eventually, multiple habs can be sent to the same site and linked together, allowing for the beginning of a permanent human settlement on the planet Mars.

As scripted, Zubrin's plan drastically lowers the amount of material that must be launched from Earth to Mars. That's a key factor to any practical plan for Mars exploration and permanent homesteading…my priority goal.

Although I differ with certain aspects of Mars Direct—favoring use of cyclers which I will discuss next—plus pre-placement of Mars habitation modules via teleoperation from Phobos—I applaud Zubrin's spirited nature.

His ideas and leadership dedication are part of a movement that is hastening the day for human settlement of Mars.

Mars Society participants sense, as I do, the untapped reservoir of individuals who value the psychology of becoming a pioneering settler, ready to jump at the opportunity to leave Earth and reside on the red planet.

History shows us that people are willing to risk their lives for great exploits of exploration. Consider the founding of Jamestown in Virginia or the Pilgrims setting foot in Plymouth, Massachusetts—these were daring one-way journeys that led to establishment of permanent settlements.

Why then, should the call of a New World Mars settlement be any different? [278]

Chapter Twenty-Eight: Cycling Superhighways to Mars and Beyond

FORMER NASA ADMINISTRATOR Tom Paine, who chaired the National Commission on Space, authored a seminal report in 1986 titled *Pioneering the Space Frontier,* stressed the need for a bridge between worlds, calling attention to the important role of cycling spaceships as a better way to gain access to Mars to avoid enormous fuel costs to accelerate and decelerate large spaceships.

The report notes that cycling spaceships permanently shuttling back and forth between the orbits of Earth and Mars would need only minor trajectory adjustments on each cycle.

So, my better-way approach for homesteading Mars is via the Aldrin Cycler developed in collaboration with James Longuski, professor of aeronautics and astronautics at Purdue, along with colleagues at JPL.

But first of all, keep in mind two terms when considering this transportation system: there are cycler trajectories and cycler vehicles.[279]

The cycler vehicles are spacious spacecraft that would provide radiation shielding and living quarters to support the safety and comfort of outbound-to-Mars and inbound-to-Earth astronaut crews.[280]

Cycler trajectories are the paths that cycler vehicles travel on. In many ways, they can be thought of as the highways on which space vehicles travel. Cycler trajectories are routes used over and over again on paths around the Sun. These trajectories are identified by using the laws of celestial mechanics—essentially Newton's laws.[281]

Newton's laws describe the relationship between the forces acting on a body and its motion due to those forces. My cycler design depends on these basic principles to create a repeatable trajectory requiring little fuel to maintain.[282]

The idea is somewhat analogous to cruise ships that drop off and take on passengers without pulling into harbor, except that cyclers don't stop when they fly by Earth.

Much like railroads and roadways that have formed transportation backbones throughout vast expanses of terrestrial wilderness, the cyclers will glide endlessly along Space Expressways within the inner Solar System. And like airlines, the strategy doesn't throw away spacecraft after reaching destinations.

Aldrin Cyclers offer a big cost-saving advantage because the massive elements need only be launched once.

Although this initial expense is significant, it pays large dividend benefits over the long-term by offering reusability along with much shorter time-of-flight requirements for astronauts.

The Aldrin Cycler uses a special trajectory that travels around the Sun, making close flybys of Earth and Mars, a trajectory that takes $2\,^1/_7$ years to complete and then repeats every succeeding $2\,^1/_7$ years. If a vehicle is launched into the Aldrin Cycler trajectory, it would continuously shuttle between the two planets virtually forever, without requiring a significant amount of propellant to keep on track.[283]

The cycler vehicle does not stop when it flies by Earth. The astronauts board a small but very speedy space taxi that catches up with the cycler. The cycler is like a bus that repeats its route over and over, but never stops. As a future space traveler, you'll have to run fast to catch up and get on the bus.[284]

But once the astronauts are on the cycler vehicle, they can relax and enjoy the ride to Mars. When they arrive at Mars, they must board a small vehicle that makes a fiery entry into the atmosphere of Mars. If the astronauts do not get off at Mars, then they will travel back to Earth, getting off $2\,^1/_7$ years after they first left Earth.[285]

By using the Aldrin Cycler trajectory, it takes less than 6 months to get to Mars. However, any astronaut not disembarking at Mars would spend 20 more months getting back to Earth.

My Purdue University associates have identified Aldrin Cycler trajectories that make a short trip—6 months— to Earth from Mars, and a long trip—20 months—to go from Earth to Mars.[286]

Therefore, a complete Earth-to-Mars human transportation system would include two cycler vehicles, one using the outbound or up-escalator trajectory to get to Mars and the other using the inbound or down-escalator trajectory.[287]

Once these cycler vehicles are built and placed in orbit about the Sun, they will continue to freely travel back and forth.[288]

As previously mentioned, some propellant will be required now and then, to keep the Aldrin Cycler going— but the cost of refueling is not prohibitive.[289]

What are the biggest challenges? The Aldrin Cycler requires very high rendezvous velocities at both Earth and Mars—typically 6 km/s (over 13,400 mph) at Earth and as high as 10 km/s (22,370 mph), or more, at Mars.

Those speeds make it very difficult for the space taxis to catch up. Think of it this way: if a bus were going 5 miles an hour, riders could easily jump on, but not if the speed was 50 miles an hour.[290]

Can anything be done about the high rendezvous speeds? Yes. My Aldrin Cycler idea has inspired the search for other Earth-Mars cycler concepts.

For example, there are low-thrust cyclers that use electrical propulsion to reduce the approach speeds. There are also four vehicle cyclers that take years to complete their trajectories. Then there are powered three-synodic-period cyclers that require three cycler vehicles. There is even a one-vehicle cycler.[291]

All of these new cyclers are spin-offs of my original Aldrin Cycler thinking, and all have much lower flyby velocities. Each has its advantages and disadvantages. As always, economics is a factor as more vehicles mean more cost. Overall, powered and low-thrust cyclers will demand advancements in propulsion technologies—but this type of progress is well within reach.[292]

Although offering no significant advantages for relatively far shorter four-day transits to the Moon and back, the sequential buildup of a full cycling network geared for maturation of lunar/Mars activities between the Earth, Moon and Mars can yield substantial savings. Such a network can form a celestial tract between worlds to support a constant flow of space science, and commerce.

So, how shall we go to Mars?

The best, most effective way is still under intensive review. But I'm confident that the Purdue/Aldrin Cycler and its offspring will continue to be an important mission design concept in the future development of an Earth-Mars transportation system for human space travel.[293]

Chapter Twenty-Nine: The New Martian Pilgrims

MY PROPOSED GOAL of permanent Mars habitation is entirely different than just putting people on the Moon. Preparations for long-term survival in harsh environments far beyond Earth lifelines pose a variety of far more difficult human adaptation and technological challenges.

The new pilgrims will require an ability to live off the land, a circumstance that 102 other adventuresome souls once faced upon leaving England for a closer New World aboard a Mayflower voyage. Martian settlers, however, will face much stiffer tests. A crop failure, for example, could bring disastrous consequences.

Much research is needed to develop and test technologies and methods to grow crops such as potatoes, beans and wheat in the thin Martian atmosphere using hydroponic and aeroponic methods that utilize on-site soil-derived water, oxygen and recycled wastes. This challenging area of study should involve scientific collaborations with Russia, China and other countries that are known to share this interest.

The particular location selected for settlement will have important resource implications. One good candidate destination will be the Cebrenia Quadrangle site located in the northeastern portion of Mars, a location that has been shown by the Mars Orbiter Laser Altimeter satellite to have water ice about 4 inches below the surface along with soil rich in silicon, iron, magnesium, Sulphur, calcium and titanium.

Means to provide for health safety will be of foremost concern. This must include exercise equipment to minimize weakening of the heart and other muscles under prolonged weightless during the voyage to Mars and the reduced one-third Earth- gravity conditions on its surface.

Extremely limited accommodations for specialized medical equipment will make access to expertise and tele-medicine support from Earth very necessary. Extensive psychological as well as physical crew selection pre-screening and post-launch monitoring will also be vital.

Since undiscovered pre-departure or subsequently developing psychological problems under isolation and other stresses can have disastrous consequences, means to continuously monitor and rapidly respond to crew mental abnormalities must be accorded a high priority.

Mental health protocols are likely to include portable on-board EEG-type monitoring and the application of Magnetic Resonance Therapy (MRT) zapping when a severe crew member behavioral disturbance is reported or remotely detected by psychological flight surgeon specialists on Earth.

Radiation shielding will be required to protect crews and equipment from dangers posed by solar flare emissions, both on the long voyage to Mars, and on its surface.

Reliability, redundancy, preventative maintenance and means to repair all life and mission-critical hardware and software systems will be essential. Some tools and replacement parts might be created onboard using 3-D printing.

Yes, inevitably, Mars settlement will invoke risks and casualties, just as other pioneering ventures have.

Unfortunately, pioneers will always pave the way with sacrifices. Over the decades, we have lost numbers of individuals—several of them close personal friends of mine—all intent on pushing the boundaries of exploration and seeking new horizons. Risk and reward are the weighing scale of exploring and taming space.

Permanent Mars settlers will be 21st-century pilgrims, pioneering a new way of life. That will indeed take a special kind of person. Instead of the traditional pilot/scientist/engineer, Martian homesteaders will be selected more for their personalities…flexible, inventive, and determined in the face of unpredictability.

In short, they will be the founders of a two-planet species.

Chapter Thirty: Charting a Bold Future

WE ARE AT AN important inflection point in human history. The decision is whether to look upwards and gain strength from vision and commitment to worthy goals beyond ourselves—beyond the here and now.

The first priority, as always, is real vision and commitment at the top.

It is time we sailed the sea of space once more with bold, expansive vision. To achieve this we need strong leaders, for sustaining a growing and momentous effort in space will require that we reject a defeatist mentality that mires us in the past, accepting the losses of jobs that lack of leadership will cause. We must set our sights higher and be prepared to sail against the wind.

Not everyone will understand this need for America to lead the world in space.

Valiant strides forward in space not only reflect our country's greatness, but summon us to make discoveries that, in turn, improve our lives on Earth. I also sense that national leadership and coming together of the American people are vital ingredients that make overcoming obstacles possible.

What does human spaceflight do for America? It reminds the American public that nothing is impossible if free people work together to accomplish great things. It captures the imagination of our youth, it fuels the American workforce and economy with high technology jobs, and it fosters peaceful and beneficial international collaborations to ensure U.S. foreign policy leadership.

Apollo 11 is a symbol of America at its best, people working together motivated by strong leaders with vision and resolve. The Gemini program served as a fundamental stepping-stone bridging between the one-man Mercury and three-person Apollo. It provided trial runs for rendezvous and docking scenarios in Earth orbit to train astronauts and ground crews.

In my frequent travels around the world, I observe with sad irony that American leadership in space is appreciated more in foreign lands than it is within our own country. Many people I meet ask why we should invest huge sums of money going to space at times when there are so many important serious problems and needs at home.

My friends, there always have been such problems and needs, and there always will be. Great nations, great people, have always faced them, confronted them, and triumphed over them. That is the bold spirit and confidence that made them great. That is the true character that defines America.

Reestablishing American Leadership

Nevertheless, another space race back to the Moon would be counterproductive. Instead, the U.S. should chart a course toward global leadership without spending taxpayer money to put and support people there. America's primary lunar focus should be to put robots on the surface for scientific, commercial, and other private sector work.

Many lessons and technologies can draw upon routine uses in other extreme environments such as deep-sea pipeline maintenance operations applying high-definition video cameras, sensors, and manipulators. The Moon will provide an excellent place to practice and perfect these technologies and operational techniques to scout out mining opportunities, pre-position and connect large equipment elements, and conduct tasks that would present dangers to surface crews.

An important American contribution can be to provide a communications infrastructure that enables surface infrastructure and operations provided by other nations to be controlled remotely from orbiting outposts in cislunar space. These services, such as means to control surface rovers and assembling lunar base habitat modules, can be offered in exchange for passenger and cargo delivery to the lunar surface and back.

A GPS navigation system will be extremely useful. Availability of a constellation of 4-5 orbiting communication relay satellites can offer blackout-free communication coverage over most of the lunar surface as well as direct line-of-sight of the Moon's far side from Earth.

America's immediate goal should be to develop new strategies, new launch vehicles and new spacecraft for years beyond…pioneering initiatives and developments on the moons and landscape of Mars, with the Earth's Moon as a technological and strategic roadway paving stone along the way. Installation of relay satellites and later, fuel depots on the surface, can greatly facilitate getting us there.

A Global Coalition

Industrialization of the Moon may fuel the way for reusable human interplanetary spacecraft, large cislunar telerobotic operations and propellant depots, and human settlements on Mars.

America can provide vital leadership to advance these futures through a global coalition of spacefaring nations that share cost burdens and benefits. Joint partners would include the U.S., Europe, Russia, Japan, and China.

The arrangement would invite participation and investments from government and private sector organizations, building upon lessons and achievements of successful NASA and Russian Space Agency commercial space privatization initiatives.

My proposed Global Low Earth Orbit (LEO) Lunar Coalition would combine mutually beneficial cost and service-sharing efforts in a manner similar to the 16-partner International Space Station arrangement, with flexibility to contract with private firms for special services.

Coalition organization and operations can also draw instructive government-to-government lessons from a successful history of international Antarctic programs. The U.S. has interacted with other nations there since 1956 with visiting exchanges of researchers involved in glaciology, biology, medicine, geology, oceanography, astronomy, and astrophysics.

Contractors and military units provide operational support at year-round stations: Palmer, the Amundson-Scot South Pole, and the main U.S. station, McMurdo. The U.S. Antarctic Program supports a peak population of about 1,600 people.

America cannot afford to throw away 50 years of accumulated human spaceflight experience and knowledge. Avoiding such a colossal waste will require top-level government and aerospace industry vision and guidance to motivate NASA and the international space community to engage in partnerships based upon clear understandings of goals, requirements, and benefits.

Practicing Tough Love

Many friends and colleagues will attest to the fact that I am frequently undiplomatic and unpolitic in offering assessments of many government space programs. A good example is in regard to the current Space Launch System (SLS), a program that was cooked up after the Constellation program was axed. I'm among some critics who have referred to it as the *Senate Launch System.*

In my view, SLS is actually little more than a resurrection and renaming of the Ares V cargo booster originally planned for the Constellation program—along with a continuation of the Orion.

Both are being developed in response to prevailing interests in leading political and industrial circles. Still, Orion can serve as an indispensable permanent emergency return vehicle or crew safe haven module for the International Space Station.

When asked to present my thoughts about this before a U.S. Senate subcommittee, my response might admittedly have been more tactful. Instead, I groused:

> *Can you believe that the law says NASA will use Heritage components to build the SLS? Do you know what that means? It's old stuff. That's not what America is about!*

At that point, one of the staffers spoke up, saying "Sir, I'm the one who wrote that into law." Rather than leaving bad enough alone, I made matters worse, responding, "Well, that's stupid!"

President G.W. Bush's vision for space exploration had problems as well. He failed to fully fund the program as he had previously promised to do, causing the rockets and spacecraft called for in the plan to slip further and further behind schedule.

Also, and even more importantly, NASA virtually eliminated technology development programs for advanced space systems.

Equally as bad, the agency raided the Earth and space sciences budgets to keep the Constellation project on track. And even as that effort fell short, the administration continued to focus on the Moon and pretty much abandon hope and preparations for human Mars exploration.

The Bush policy focused on U.S. astronauts returning to the Moon as early as 2020. The central priority emphasized gaining benefits afforded by access to lunar surface resources that can be processed into rocket fuel and breathable air.

Rolling out his policy, the president said:

> *We can use our time on the Moon to develop and test new approaches and technologies and systems that will allow us to function in other, more challenging, environments. The Moon is a logical step toward further progress and achievement.*

The influential Obama administration panel headed by my friend Norm Augustine properly recognized that Constellation couldn't be executed without very substantial increases in funding, and that the path established by the Bush administration was not sustainable.

It is vital that we take lessons from previous mistakes and not repeat them. In retrospect, it was a big mistake for NASA not to design the Space Shuttle as a fully reusable two-stage system with a recoverable liquid chemical first stage booster rather than single-use solid rocket motors, a technology that keeps popping out of a casket.

I favor separate launch vehicles for people and cargo to overcome a fundamental error that bundles different safety problems together and unnecessarily boosts costs of access to space.

In 2003, former NASA engineer Hubert Davis and I formed a rocket design company, Starcraft Boosters, Inc., to implement a two-stage Star Booster fly-back launcher and lander system. The patented design is essentially a hollow aircraft-type airframe into which a booster rocket propulsion module—such as a liquid-fueled Atlas V, Delta IV, or Russian Zenit is inserted…much like putting replaceable liquid-fuel cartridges in a modern fountain pen.

The aim is of this Star Eagle spaceplane was to provide airline-type operations with high-reliability, quick turnaround, and a large passenger capacity.

I also urge that NASA get out of the political climate change business and concentrate on doing what it was set up to do and what taxpayers pay it to do…reestablish American space leadership and real progress with well-defined step-by-step pathways to Mars.

Let's realize that the climate has been changing for billions of years, alternating between warmer and cooler, and for better and worse. That is what climate does. If it didn't, no one would have found it necessary to invent the word.

I'm not in favor of just taking short-term isolated situations and depleting our resources to keep our climate the way it is today. I'm also not of the school that we are causing a climate threat. The world does what it does, sometimes for the better; sometimes not. We should get use to that and plan accordingly.

In any case, going back to the Moon and replaying the glory of Apollo is not visionary. It will not advance the cause of American space leadership or inspire the support and enthusiasm of the public and the next generation into space explorers.

While Apollo is a symbol of what a great nation can achieve, the United States can regain a strong leadership role both in low Earth orbit and on the Moon without sending NASA astronauts back to the lunar surface with taxpayer money. They have better places to go.

Like the Apollo predecessor, returning to the Moon with a get-there-quick-and-leave-again program will prove to be just another dead end littered with broken spacecraft, broken dreams, and broken spacecraft.

I'm not suggesting that America abandon the Moon completely…but only that we forgo it as a destination priority, focusing instead upon assisting rather than competing with other countries.

American leadership in lunar exploration and development should be directed toward guiding other spacefaring nations in preparation for continuing activities that will lead to permanent human settlement of Mars. Earth and the Moon aren't the only worlds for us anymore.

While the lunar surface can be used to develop advanced technologies, it is a poor location for homesteading…a lifeless, barren world of stark desolation. Working with international partners we can use the Moon as a testbed for tools, equipment and operations that will be needed for our next major destination…Mars.

NASA's lunar participation should focus upon telerobotics controlled from a distance—both on the near side and far side—like, for example, assembling lunar exploration modules constructed and landed by partner countries. We should not, however, train and land our astronauts there. I maintain that any taxpayer money should instead be spent on advancing preparations for occupation and performance on Mars.

America should lead a consortium of international partners, including Europe, Russia, Japan, and China, with the necessary planning, coordinated technology development, and execution of manned missions to the Moon.

Through validation of habitation and other systems on the Moon, this consortium can prepare similar equipment for landing and operation at selected Mars base locations.

China is inviting international collaboration for commercial Earth orbit activities and is open to future joint Moon missions, but is also prepared to go it alone.

Russia and China have announced that they are preparing plans to build a joint International Lunar Research Station (ILRS) intended to enable crewed visits by 2036. They have announced that ILRS will invite collaborations with other countries.

America needs to develop new strategies, technologies and spacecraft to bring us to the threshold of Mars by way of progressive missions to comets, asteroids, Mars' moons, and a permanent Mars settlement.

In doing so, we fly by comets and asteroids and sweep their surfaces to discover what the building blocks of the universe are made of…step-by-step…just as Mercury and Gemini made Apollo possible. We move deeper into space to land on Phobos or Deimos, all in prelude to the red planet itself.

We can create Space Expressways of beautiful simplicity that use gravitational forces to route cyclers back and forth between two worlds—Earth and Mars—and points in between. The first transit from Earth to Mars will be an unmanned demonstration.

Later transits will deliver pioneering homesteaders who will stay. Every time a cycler swings past Earth, it will be met by a supply ferry and boarded by a fresh group of pioneering crews destined for the Mars surface or one of its moons.

New World of Opportunity

Mars represents a new world of opportunity and discovery. Scientific and public interest in the planet has grown since 1960 telescope-driven observations have since been augmented by voyages of numbers of automated spacecraft sent there by multiple nations.

While robotic exploration of Mars has yielded tantalizing clues about what was once a water-soaked planet and has revealed frozen water still trapped below the surface, the best way to study Mars is with the two hands, two eyes and two ears of a geologist, first at a moon orbiting the planet…and then on the surface.

Humans trump machines in speed, efficiency, nimbleness, and the dexterity to go places and do things. We have the innate smarts, ingenuity, and adaptability to evaluate and respond in real-time situations…to improvise, and to prevail over surprises.

Recent and ongoing robotic exploration of Mars is providing a window on a world that can be home for new generations of colonists.

This is far different than Apollo expeditions to the Moon where voyagers do some experiments, plant a flag, and claim success. Having them go to the red planet and repeat this, in my view, is senseless. Since great distance between Mars and Earth makes a feasible return window very narrow, it makes far more sense to transport people there who plan to stay.

So yes, I suggest that going to Mars means preparing for permanence on the planet. This, of course, is a big high-level policy matter. Any legislative branch we have today will insist on returning crews as rapidly as possible even if the most economical and useful period is one and one-half years and it means that Mars will be vacant for more than a year before the next crew arrives on a similar mission. In addition, the first tragedy is likely to result in cancellation of the program for a century.

Success at Mars cannot stop with one-shot forays to the surface. Yet I'm painfully aware that my views regarding permanent human presence are considered by many to be very radical, if not simply controversial and a political hard sell. I'm often asked in various ways: "Lifetime trips to Mars? That's a big pill to swallow!"

Yet from my viewpoint, and with certain personal experience in such matters, let me pose a few questions to these skeptics.

What are you going to do with astronauts who first reach the surface of Mars and then turn around and rocket back homeward? After investing billions and billions of dollars of world assets in getting them there, why have them turn around and rocket back homeward?

And why spend billions and billions of dollars more bringing them back if they don't wish to return? Their value in remaining of Mars together with previous crews represents a great return on investment.

Having them repeat their voyages, in my view, is dim-witted. Why not allow them to stay there? Did the pilgrims on the Mayflower sit around Plymouth Rock waiting for a return trip? They came to settle. And that's what we should be doing on Mars.

When you go to Mars, you need to have made the decision to go there permanently. The more people you have there, the more it can become a sustaining environment. Except for very rare exceptions, the people who go to Mars should be prepared to remain.

A NASA report observes that a strong motivating factor for the exploration of Mars is the search for extraterrestrial life.

Maybe what we will discover there is us.

Maybe what we will discover are human potentials we can now only dream about.

Risking Failure is Always a Necessity

As I reflected in my earlier book *No Dream is Too High*, if you want to do something significant, something noble, something that perhaps has never been done before, you must be willing to fail. And don't be surprised or devastated if you do. Untold numbers of people have experienced major failures and have recovered to become not only successful, but also as better, stronger people.

Failure is not a sign of weakness. It is evidence that you are alive and accepting worthwhile risks. Successful innovators and doers who conceive things and make them happen combine awareness of failure risks of worthwhile enterprises with the willingness to take them. They are patient, resilient and don't quit.

Instead, they experiment, often fail, learn more, and start over. Great companies like Google and Apple provide a culture that empowers employees to stretch their curiosity and creativity to explore impossible dreams that often don't work out.

As Albert von Szent-Gyorgi, the Hungarian Nobel Prize-winning physiologist who first discovered the benefits of vitamin C, was fond of saying:

> *Discovery lies in seeing what everyone sees, but thinking what no one else has thought.*

And as an old adage goes: Can't never could. No never will. Success comes in cans.

Time for New Vision and Commitment

In Korea, we knew we were really fighting the Soviets as well as the North Koreans, and a strong sense of competition on our part carried into the space race. We were determined not to let the *Ruskies* beat us in Korea, and we certainly weren't going to let them get the upper hand in space.

That Cold War space race during the 1960s and '70s to outperform the former Soviet Union has reached the finish line. A second race to the Moon would be a waste of precious resources. It will offer no unique American glory or payoffs in either commercial or scientific terms. It's high time now to raise our vision and commitments to loftier, more far-reaching goals in global cooperation.

Apollo was all about a get-there-in-a hurry straightforward space race strategy and don't waste time developing reusability. That chapter in the space exploration history books is closed. Instead, I urge that all spacefaring nations join a unified international effort to explore and utilize the Moon through a partnership that involves commercial enterprises and other nations.

America must once again dare to pursue big dreams. It is my great hope that a new generation of leaders and doers will once again boldly venture where no one has gone before. Our Apollo days were a time when we did bold things, achieving leadership.

Now is our time to be bold again in space.

Chapter Thirty-One: Addressing Threats from Above

I STRONGLY SUPPORT former President Trump's establishment of U.S. Space Force, recognizing that while space represents frontier of great peaceful global opportunity and progress, it also poses an environment of latent manmade and natural security threats. Planetary defense of home planet Earth means getting to know the enemy— and I am not exclusively talking about down-to-Earth squabbles between nations. I'm highlighting here a celestial fear factor stemming from asteroids and comets.

NEOs have been nudged by the gravitational attraction of nearby planets into orbits that allow them to enter Earth's solar system neighborhood. We should therefore learn more about these extra-terrestrial wanderers in both scientific and practical terms.[294] So, let's look at the exploration, science, development, commerce, and security pieces that are tied to NEOs.

Number one is that near-Earth objects have thumped our world over the ages, and assuredly will in the future. NEOs can shake up but also shape our life-sustaining ecosystem. To assure the survival and guarantee the movement of humanity into space, I feel it is vital we come to terms with NEOs that may have Earth within their crosshairs.

Doing so harnesses the technological muscle to not only encounter but also counter these objects, and it also allows us to use space objects as a resource and exploration stepping-stones to Mars, thereby helping to extend the human presence into space.[295]

It turns out that smaller airbursters are the more disconcerting sky-slamming flotsam from space. For example, the 1908 Tunguska event is a saga in which a rocky impactor detonated over remote Siberian real estate, knocking down about 500,000 acres of forest.[296]

Because asteroids approach Earth statistically more often than larger ones, efforts to detect smaller NEOs would appear to be in order. It is estimated that these smaller objects could impact Earth on average every 2 to 12 years. In October 2009, a fireball blast in daylight was observed and recorded over an island region of Indonesia. That atmospheric entry of a small asteroid, perhaps just 33 feet across, rocked their world with a projected energy release of about 50 kilotons, equal to some 110,000 pounds (50,000 kg) of TNT explosive.[297]

Let's face facts. We live in a cosmic shooting gallery. Ways to defend ourselves from NEOs demand careful study.[298]

Chapter Thirty-Two: Some Life Lessons and Recommendations

I WAS DEEPLY saddened by the passing of my good friends, and space exploration companions, Neil Armstrong in 2012, and Mike Collins in 2021.

As Neil, Mike, and I trained together for our momentous Apollo 11 expedition, we knew of the technical challenges we faced as well as of the magnitude and weighty implications of that historic journey.

We will now always be connected as the crew of the Apollo 11 mission to the Moon in 1969. Yet for the many millions who witnessed that remarkable achievement for humankind, we were not alone. An estimated 600 million people back on Earth, at that time the largest television audience in history, watched Neil and I walk on the Moon.[299]

Whenever I gaze at the Moon, I feel like I'm in a time machine. I am back to that precious pinpoint of time more than a half-century ago when Neil and I stood on the forbidding, yet beautiful, Sea of Tranquility.

We both looked upward at our shining, blue planet Earth, poised in the darkness of space. I now know that even though we were farther away from Earth than two humans had ever been, we were simply the spearhead of a community of participants. Virtually the entire world took that unforgettable journey with us.[300]

With Neil's death, I was joined by many millions of others from around the world in a global mourning for the passing of a true American hero and the best pilot I ever knew. It had never come to my mind that our Apollo 11 mission commander might be the first of us to pass.[301]

My friend Neil took the small step but giant leap that changed the world, and he will forever be remembered as the person who represented a seminal moment in human history.

I had truly hoped that on July 20, 2019, Neil, Mike, and I would be commemorating the 50th anniversary of our Moon landing. Regrettably, this was not to be.

Surely, if we had all been together, we would have collectively supported the continued expansion of humanity into space. Our small mission that was Apollo 11 helps make that possible. But like our fellow citizens and people from around world, we all will miss these space pioneers.[302]

Neil did not see Apollo 11 as an ending. Rather, he saw our touchdown at Tranquility Base as a first small step for humankind into the cosmos. He was truly a gifted engineer, consummate astronaut, and leader.

Yes, he was soft-spoken and reserved, advocating quietly for space exploration from behind the scenes. He didn't seek fame or honor for the work that he knew so many others had done to make our Moon landing achievable.[303]

May Neil's vision for our human destiny in space be his legacy. As he once observed, there are still places to go beyond belief.

We both expressed that sentiment during periodic visits to the White House, where we discussed U.S. space policy with a succession of presidents.

Conversation in some cases turned to where the next step into the future should lie: return to the Moon or on to Mars? For me, Mars.

Neil disagreed. He thought that the Moon had more to teach us before we pressed onward to other challenges. Still, while we disagreed at times on that next destination and how best to get there, we were both resolute and shared a common belief: America must lead in space.[304]

Neil and Mike's passings were also times to reflect on those who gave their lives in pursuit of making real the dream of space exploration: the astronauts of Apollo 1 and of the space shuttle orbiters Challenger and Columbia.

We can honor them all—and the U.S. president who set in motion the Moon-landing challenge before the country—by renewing our commitment and resolve to space exploration and pursue it with the same fortitude and durable commitment to excellence that was personified by Neil and Mike.[305]

My call is to the next generation of space explorers and their leaders. It is now time to continue that journey, outward past the Moon.

The three of us on Apollo 11 traversed the blackness and vacuum of space to win a peaceful race with a very capable competitor, the former Soviet Union.

Apollo 11 was, at its core, about leadership. A young American president challenged himself—and all of us—to think daringly and not withdraw from a shared vision of what we could do in space.

The path that John F. Kennedy motivated us to choose was, indeed, not easy. In truth, it was very hard and audacious in scope. But it served the betterment of America, and ultimately contributed to ending the Cold War.[306]

It was an honor and privilege to be a part of the Apollo 11 mission and share in that pinnacle moment within the U.S. space program.

There are those who look back at that time and ask, "What did it mean that America was first on the Moon?" The right question to ask, however, is, "What can America do now to build upon that accomplishment decades ago?"[307]

Apollo 11 was rooted in exploration, about taking risks for great rewards in science and engineering, about setting an ambitious goal before the world—and then finding the political will and national means to achieve it.

Even today the voyages of Apollo seem incredibly bold. Looking back at that time, we are continually stirred by the enormity of the endeavor, one that was energized by the teaming efforts of people from all walks of life, from industries big and small, who worked in tandem to attain a long-term goal of magnificent achievement.[308]

The crew of Apollo 11 was backed by hundreds of thousands of American workers, the greatest can-do team ever assembled on the face of Earth.

That team was composed of scientists and engineers, metallurgists and meteorologists, policymakers and flight directors, navigators and suit testers, and those on the shop floor, such as the seamstresses who stitched the 21 layers of fabric for each custom-tailored space suit.

They devoted their lives and professional energies, minds, and hearts, to our mission and to the other Apollo expeditions. Those Americans embraced commitment and quality to surmount the unknowns with us.[309]

All of these lessons are worth learning anew today. Yes, we live in difficult times. We face these challenges together.

I believe that valiant strides forward in space not only reflect our country's greatness, but summon us to make discoveries that, in turn, improve our lives on Earth.

I also sense that national leadership and a coming together of the American people are the ingredients that make overcoming obstacles possible. Apollo 11 is a symbol of what a great nation—and a great people—can do if we work hard, work together, motivated by strong leaders with vision and resolve.[310]

What Apollo 11 means to us today is realizing the dream of exploration by way of determination—and it is a message we need to carry forward into our future.

My vision for our space future is founded on the Apollo tradition. But this time, there is no Moon race. Rather, I see the Moon as a true stepping-stone to more stimulating and habitable destinations.

The Moon should act as a new global commons for all nations as we venture outward to Mars for America's future. It is not outside our reach.[311]

That future is already being cultivated as U.S. space entrepreneurs are opening up the space ways to tourism for hundreds of ordinary individuals.

It is a future in which we build upon great private sector commercial accomplishments to expand travel to Earth's orbit aboard reusable spacecraft supporting a wide variety of space duties.

It is a future of interplanetary transport that can be combined with Orion-type crew vehicles for missions that cycle back and forth between Earth and the moon.

It is a future of crewed habitats and robotic spacecraft, relay satellites and refueling depots orbiting around Earth-Moon L-1 and L-2 positions.

This is nearly a snapshot of what is predictable.

Who knows what the future may bring, what's right around the technological corner or what new revelation in physics, quantum theory for example, is yet to be found?

Propulsion via gravity waves, space elevators on the Moon, satellite power beaming from point to point in space, our first contact with extraterrestrial intelligence?

By implementing a step-by-step vision—just as we did with the single-seater Mercury capsule and two-person Gemini spacecraft that made Apollo possible—we will plunge deeper and deeper outward.

On the agenda of solid stepping-stones in space exploration: multination and commercial use of our neighboring Moon, several human landings on Phobos, the inner moon of Mars. Those exploits are prelude to our historical and milestone-making commitment to homestead the red planet itself.[312]

If collectively we have the vision, determination, support, and political will—and Apollo clearly showed us that these elements can be tied together—then these gallant missions of exploration are within our grasp.[313]

America, do you still dream great dreams? Do you still believe in yourself? Are you ready for a great national challenge?

I call upon our next generation of space explorers—and our political leaders—to give an affirmative answer: Yes![314]

I have a message in a time bottle for the candidate who wins the 2024 election for the U.S. presidency:

> *I believe this nation should commit itself, within two decades, to commencing American permanence on the planet Mars.*

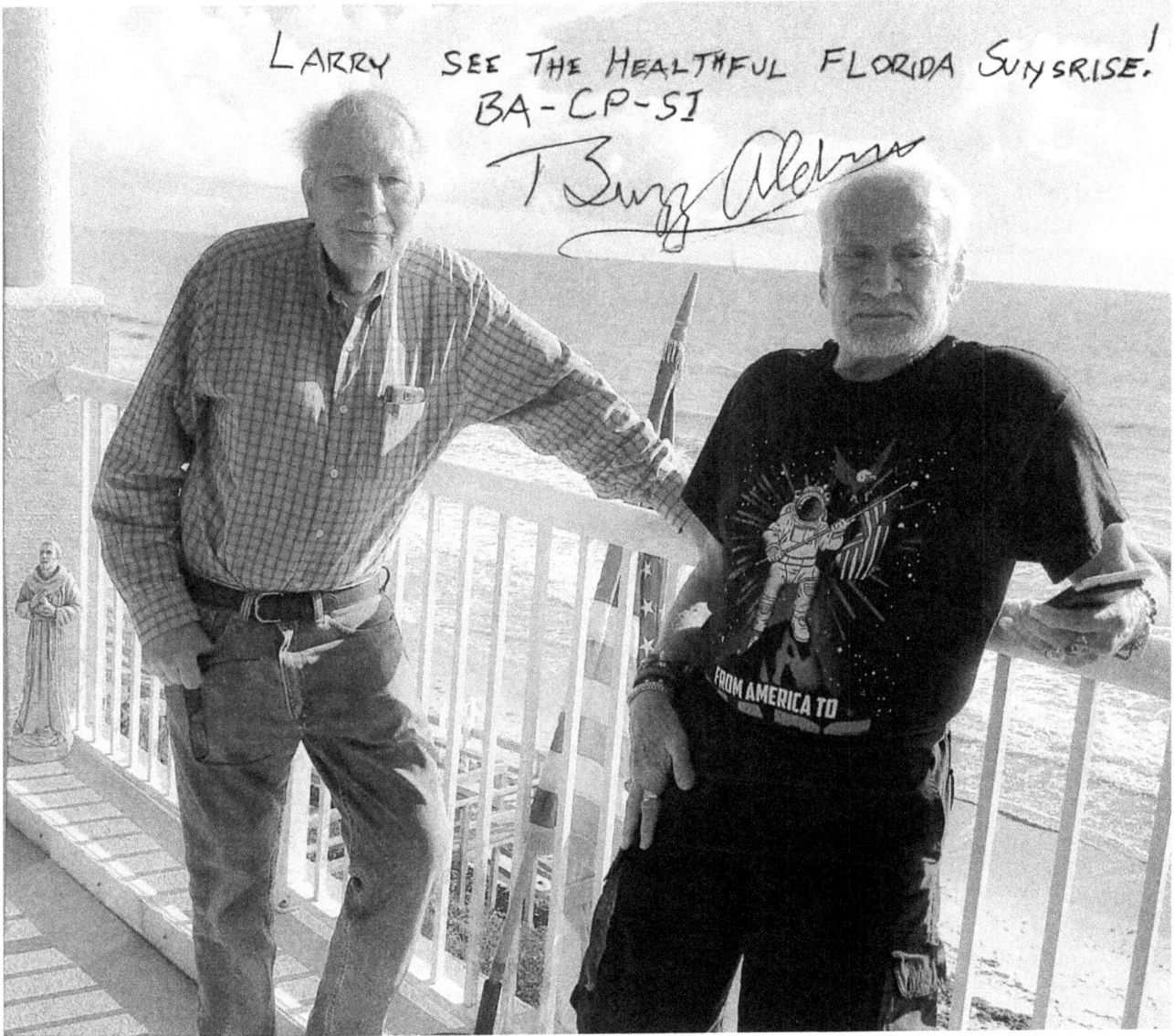

Index

Endnotes

[1] Korolev, Sergei P., *The Practical Significance of Konstantin Tsilokovsky's Proposals in the Field of Rocketry,* paper presented at a meeting commemorating Tsiolkovsky's 100th birthday, Sept. 17, 1957, USSR Academy of Sciences, Moscow, reprinted in English in *History of the USSR;* New Research, 5, Social Sciences Today, Moscow, 1986, p.52.

[2] Ibid.

[3] Ibid.

[4] Ibid.

[5] Golovanov, Yaroslav, *Sergie Korolev: The Apprenticeship of a Space Pioneer*, Moscow: Mir, 1975.

[6] Keldysh, M.V., editor, and Vetrov, G.S., compiler, *Creative Legacy of Academician Sergei Pavlovich Korolev*, (in Russian), Moscow, Nauka, 1980, p.54.

[7] Hartford, James, Korolev: How One Man Masterminded the Soviet Drive to Beat America to the Moon, Wiley & Sons, Inc., 1997, pp. 49-51.

[8] Ibid.

[9] Goddard, Esther C., and Pendray, G. Edward, *The Papers of Robert H. Goddard, Vol.II,* 1925-1937, New York: McGraw-Hill, 1970, pp.588-89.

[10] Ibid.

[11] Goddard, Robert H., *Report Concerning Further Developments,* March 1920, The Smithsonian Institution Archives.

[12] *Topics of the Times,* New York Times, January 13, 1920.

[13] Leo Nutz; Elmer Wild, December 28, 1989, Oberth-museum.org

[14] http://biography.com/people/wernher-von-braun-9224912.

[15] *Recollections of Childhood: Early Experiences in Rocketry as Told by Wernher von Braun*, 1963, MSFC History Office, NASA Marshall Space Flight Center.

[16] *Excerpts from 'Power to Explore',* MSFC History Office, NASA Marshall Space Flight Center.

[17] Middlebrook, Martin, *The Peenemunde Raid: The Night of 17-18 August, 1943, 1982,* New York, Bobs-Merrill, p.222. ISBN 0-672-527759-6.

[18] Neufeld, Michael J., Wernher von Braun: Dreamer of Space, Engineer of War, 2008, Vintage, p. 184.

[19] Chertok interview in Izvestia, Nos. 54, 55, 56, 57, 58, March 4-9, 1992.

[20] Hartford, James, Korolev: How One Man Masterminded the Soviet Drive to Beat America to the Moon, Wiley & Sons, Inc., 1997, p. 64.

[21] Chertok interview in Izvestia, Nos. 54, 55, 56, 57, 58, March 4-9, 1992.

[22] Scientific Intelligence Research Aid #74, Central Intelligence Agency, Washington, D.C., *Scientific Research Institute and Experimental Factory 88 for Guided Missile Development, Moskva/Kalingrad,* p.2, March 4, 1960.

[23] As reported in Hartford, James, Korolev: How One Man Masterminded the Soviet Drive to Beat America to the Moon, Wiley & Sons, Inc., 1997, p. 68:

[24] Chertok interview in Izvestia, Nos. 54, 55, 56, 57, 58, March 4-9, 1992.

[25] Chertok interview in *Nezavisimaya Gazeta*, August 19, 1993, as translated in JPRS-USP-93-005, October 5, 1993, p.30, Washington, D.C.

[26] Ordway, Frederick I. III., and Sharpe, Mitchell R., *The Rocket Team*, New York, Thomas Y. Crowell. 1979, pp.251-52.

[27] Tarasenko, Maxim, *Evolution of the Soviet Space Industry,* IAA-95-IAA.2.1.01, 46th International Astronautical Congress, Oslo, Oct.2-6, 1995.

[28] As reported in an interview with James Hartford in his book *Korolev: How One Man Masterminded the Soviet Drive to Beat America to the Moon*, Wiley & Sons, Inc., 1997, p. 79.

[29] Ivanovsky interview with James Hartford on Jan. 26, 1993 in his book *Korolev: How One Man Masterminded the Soviet Drive to Beat America to the Moon*, Wiley & Sons, Inc., 1997, p. 80.

[30] Chertok interview in *Nezavisimaya Gazeta*, August 19, 1993, as translated in JPRS-USP-93-005, October 5, 1993, p.30, Washington, D.C.

[31] As reported in Hartford, James, Korolev: How One Man Masterminded the Soviet Drive to Beat America to the Moon, Wiley & Sons, Inc., 1997, pp.81-83.

[32] As reported in Hartford, James, Korolev: How One Man Masterminded the Soviet Drive to Beat America to the Moon, Wiley & Sons, Inc., 1997, p. 85.

[33] Scientific Intelligence Research Aid #74, Central Intelligence Agency, Washington, D.C., *Scientific Research Institute and Experimental Factory 88 for Guided Missile Development,* Moskva/Kalingrad, p.2, March 4, 1960.

[34] As reported in Hartford, James, Korolev: How One Man Masterminded the Soviet Drive to Beat America to the Moon, Wiley & Sons, Inc., 1997, p. 87.

[35] Myers, Dale, *The Navaho Cruise Missile–A Burst of Technology,* Acta Astronautica, Vol. 26, No. 8, 1992, pp.741-48.

[36] *Pravda,* November 29, 1977.

[37] *Tass,* September 8, 1958.

[38] National Intelligence Estimate, *The Russian Space Program,* Dec.5, 1962, p.3, as reported in Hartford, James, *Korolev: How One Man Masterminded the Soviet Drive to Beat America to the Moon*, Wiley & Sons, Inc., 1997, p. 191.

[39] Day, Wayne, *Corona: America's First Spy Satellite Program,* Quest, Grand Rapids, MI: Cspace Press, Summer 1995, pp. 4-21; McDowell, Jonathan, *US Reconnaissance Satellite Programs,* pp. 22-33; and McDonald, Robert A., *Corona Success for Space Reconnaissance,* PE & RS Photogrammetric Engineering and Remote Sensing, June 1995, pp.689-720.

[40] *Le Figaro,* Paris, October 7, 1957, pp.4-5.

[41] Stuhlinger, Ernst, and Ordway, Frederick I.III, *Wernher von Braun: Crusader for Space,* Malabar FL: Kreiger Publishing, 1994, pp. 123-31.

[42] Keldysh and Vetrov, Creative Legacy…, p. 343, and Tarasenko, Maxim, *Military Aspects of Soviet Cosmonautics*, Moscow; Nikol 1992, p. 16, as reported in Hartford, James, *Korolev: How One Man Masterminded the Soviet Drive to Beat America to the Moon*, Wiley & Sons, Inc., 1997, p. 123.

[43] *Scientific Uses of Earth Satellites,* University of Michigan Press, January 27, 1956.

[44] Rebrov.M., The Way it Was: The Difficult Fate of the N-1 Project, Krasnaya Zvezda (Russian), Jan. 15, 1990, No.11, p.4, as reported in Hartford, James, Korolev: How One Man Masterminded the Soviet Drive to Beat America to the Moon, Wiley & Sons, Inc., 1997, pp. 246-47.

[45] Kelydysh and Vetrov, *Creative Legacy….,* pp. 400-4, as reported in Hartford, James, *Korolev: How One Man Masterminded the Soviet Drive to Beat America to the Moon*, Wiley & Sons, Inc., 1997, p. 139.

[46] Hartford, James, Korolev: How One Man Masterminded the Soviet Drive to Beat America to the Moon, Wiley & Sons, Inc., 1997, p. 141.

[47] Glennan, T. Keith, *The Birth of NASA,* Washington, D.C.: NASA History Office, 1993, p.31, as reported by Hartford, James, *Korolev: How One Man Masterminded the Soviet Drive to Beat America to the Moon,* Wiley & Sons, Inc., 1997, p. 147.

[48] *Novosti Kosmonavtiki* No. 26/18-31 Dec., 1993, p. 46, as reported by Hartford, James, *Korolev: How One Man Masterminded the Soviet Drive to Beat America to the Moon,* Wiley & Sons, Inc., 1997, p. 149.

[49] Bryushinin, V.M., *Breakthrough Into the Cosmos", in* Russian, Moscow, Veles Ltd., 1994, p.105, as reported by Hartford, James, *Korolev: How One Man Masterminded the Soviet Drive to Beat America to the Moon,* Wiley & Sons, Inc., 1997, p. 152.

[50] Hartford, James, Korolev: How One Man Masterminded the Soviet Drive to Beat America to the Moon, Wiley & Sons, Inc., 1997, p. 132.

[51] Low interview at NASA Headquarters, Washington, D.C., John F. Kennedy Oral History Project, May 1, 1964.

[52] Hartford, James, Korolev: How One Man Masterminded the Soviet Drive to Beat America to the Moon, Wiley & Sons, Inc., 1997, p.3.

[53] BBC News, The First Spacewalk: How the First Human to Take Steps in Outer Space Nearly Didn't Return to Earth, http://www.bbc.co.uk/news/special/2014/newsspec_9035/index.html.

[54] Leonov interview with James Hartford, Moscow, Dec. 11, 1991, as reported in Hartford, James, *Korolev: How One Man Masterminded the Soviet Drive to Beat America to the Moon,* Wiley & Sons, Inc., 1997, pp.185-86.

[55] Glennan, T. Keith, *The Birth of NASA,* Washington, D.C.: NASA History Office, 1993, p. 13.

[56] Logsdon, John M., *The Decision to Go to the Moon,* Cambridge, MA: MIT Press, 1970, p. 106.

[57] Memorandum from John F. Kennedy to Lyndon Johnson, Apr 20, 1961, John F. Kennedy Library, Boston, MA.

[58] Von Braun, Wernher, Memo to Vice President Johnson, April 29, 1961, NASA Historical Archives.

[59] Kennedy address to the 18th General Assembly of the UN, Sept. 20, 1963, JFK Library.

[60] Bundy, McGeorge, memo to President Kennedy, Sept.18, 1963, JFK Library.

[61] Kennedy, John F., speech prepared for delivery to the Dallas Citizens Council, Nov. 22, 1963, JFK Library.

[62] "The Apollo Soyuz Mission 1975", NASA YouTube, May 31, 2015.

[63] Battaglia, Deborah, *Arresting Hospitality: The Case of the Handshake in Space,* Journal of the Royal Anthropological Institute, vol. 18 issue, 1 June 2012, pp.S76-S89.

[64] Mir Space Station, NASA

www.history.nasa.gov/SP-4225/mir/mir.htm.

[65] Mir Space Station, NASA

www.history.nasa.gov/SP-4225/mir/mir.htm.

[66] NASA Manned Spacecraft Center and Marshall Space Flight Center, *Study of Integral Launch and Reentry System,* RFP MSC BG721-28-96C and RFP MSFC 1-7-21-00020, Oct. 30, 1968, copy in Johnson Space Center Archives.

[67] Rowland White, *Into the Black,* March, 2016, Bantam Press, Transworld Publishers, Ltd.,

https://www.bookdepository.com/Into-Black-Rowland-White/9780593064368.

[68] Coauthor Larry Bell was an original Space Industries, Inc. co-founder along with Max Faget, Guillermo Trotti, and James Calaway.

[69] Jeff Foust, *Griffin's Commercialization Legacy,* SpaceNews, The Space Review, December 8, 2008.

[70] Report of Presidential Commission on the Space Shuttle Challenger Accident, Washington, D.C.: U.S. Government Printing Office, June 6, 1986.

[71] Williamson, Ray A., Developing *the Space Shuttle: Chapter Two; Early Concepts of a Reusable Launch Vehicle*, pp. 184-85, history.nasa.gov/sts1/pdfs/explore.pdf.

[72] *Back to Space!,* Aviation Week & Space Technology, October 3, 1988, p.7.

[73] U.S. Congress, Office of Technology Assessment, *Round Trip to Orbit: Human Spaceflight Alternatives,* Washington, DC: U.S. Government Printing Office, August 1989, pp. 6, 25.

[74] *Columbia Disaster: What Happened, What NASA Learned,* Elizabeth Howell, Space.com, February I, 2013.

[75] Efraim Akim interview as reported in Hartford, James, *Korolev: How One Man Masterminded the Soviet Drive to Beat America to the Moon*, Wiley & Sons, Inc., 1997, p.314.

[76] Boris Gubanov interview as reported in Hartford, James, *Korolev: How One Man Masterminded the Soviet Drive to Beat America to the Moon*, Wiley & Sons, Inc., 1997, p.315.

[77] Aerospace America, *Challenger's Legacy*, January 2016, pp. 21-27.

[78] Pathways to Exploration: Rationales and Approaches for US program for Human Space Exploration, 2014, The National Academies Press, http://www.nap.edu/read/18801/chapter/1.

[79] *Why the Moon?,* NASA Website, December 4, 2006.

[80] White House Office of Science and Technology Policy (OSTP), May 7, 2009,

http://www.ostp.gov/galleries/press_release_files?NASA%20Review.pdf.

[81] Review of US Human Space Flight Plans Committee: Augustine; Chyba; Kennel; Bejmuk; Crawley; Lyles; Chiao; Greason; Ride, *Seeking a Human Spaceflight Program Worthy of A Great Nation*, Final Report, NASA.

[82] William Harwood, *NASA Commits to $7 Billion SLS Development,* CBS News, August 27, 2014, Space News.

[83] Larry Bell, A Discussion with Astronaut Bonnie Dunbar: Re-Engineering America's Space Leadership, Forbes.com, August 20, 2013.

[84] John C. Abell, Sept. 9, 1982, *3-2-1...Liftoff! The First Private Rocket Launch*, September 9, 2009, Wired.com, http://www.wired.com/2009/09/dayintech0909privaterocket/.

[85] James Knauf, *Freedom from Russian Rocket Engines*, Aerospaceamerica.org ANALYSIS, September 2016.

[86] https://www.blueorigin.com/engines.

[87] https://en.wikipedia.org/wiki/SpaceX_rocket_engines.

[88] SPACENEWS, *5 Companies to Watch in 2016,* November 16, 2015, http://spacenews.com/5-space -companies-to-watch-in-2016.

[89] Doug Mohney, *SpaceX Plans to Test Reusable Suborbital VTVL Rocket in Texas,* September 26, 2011, Satellite Spotlight, http://satellite.tmcnet.com/topics/satellite/articles/222324-spacex-plans-test-reusable-suborbital-vtvl-rocket-texas.htm.

[90] Stephen Clark, Reusable Rocket Prototype almost Ready for First Liftoff, Spaceflight Now, July 9, 2011, http://www.spaceflightnow.com/news/n1207/10grasshopper/.

[91] Lindsey Clark, *SpaceX Moving Quickly Towards Fly-back First Stage*, NewSpace Watch, March 3, 2013, http://www.spaceflightnow.com/news/n1207/10grasshopper/.

[92] Stephen Clark, *Sweet Success at Last for Falcon 1 Rocket,* September 28, 2008, Spaceflight Now, http://www.bing.com/search?q=Sweet+Success+at+Last+for+Falcon+1+Rocket&form=PRUSEN&mkt=en-us&refig=ec79662bc6ec441caf29c47d79872e0d&pq=%22spacex+&sc=8-8&sp=-1&qs=n&sk=&cvid=ec79662bc6ec441caf29c47d79872e0d.

[93] SpaceX Capsule Docks at International Space Station, May 25, 2012, Space.com, http://www.space.com/17998-spacex-s-cargo-capsule-docks-with-international-space-station-video.html.

[94] https://en.wikipedia.org/wiki/List_of_Falcon_9_and_Falcon_Heavy_launches.

[95] https://en.wikipedia.org/wiki/Starlink.

[96] Gwynne Shotwell Comments at Commercial Space Transportation Conference, February 3, 2016.

[97] NASA Website, NASA Awards Commercial Space Station Resupply Contracts, http://www.nasa.gov/home/hqnews/2008/dec/HQ_C08-069_ISS_Resupply.html.

[98] Stephen Clark, *SpaceX Hopes to Raise Launch Tempo after Space Station Flight*, March 20, 2016, Spaceflight Now, http://www.bing.com/news/search?q=SpaceX+Hopes+To+Raise+Launch+Tempo+After+Space+Station+Flight&qpvt=SpaceX+hopes+to+raise+launch+tempo+after+space+station+flight&FORM=EWRE.

[99] W.J Hennigan, Boeing, *SpaceX Big Winners in NASA Competition for New Spacecraft*, August 3, 2012, Seattle Times, http://www.seattletimes.com/business/boeing-spacex-big-winners-in-nasa-contest-for-new-spacecraft/.

[100] NASA Press Release, NASA Announces Next Steps In Effort To Launch Americans From U.S. Soil, August 3, 2012, http://www.nasa.gov/home/hqnews/2012/aug/HQ_12-263_CCiCAP_Awards.html.

[101] https://www.theverge.com/2019/3/7/18254549/spacex-crew-dragon-iss-nasa-landing-parachutes-splashdown https://www.cnet.com/news/spacex-crew-dragon-splashdown-see-nasa-astronauts-return-to-earth/.

[102] *Elon's SpaceX Tour—Engines,* November 11, 2010, YouTube https://www.youtube.com/watch?v=OCc2F8KccD4

[103] Steve Jurvetson, *SpaceX and Daring to Think Big,* January 28, 2015, Draper TV, YouTube, http://www.drapertv.com/videos/52

[104] Jonathan Amos, *SpaceX Launches SES Commercial TV Satellite for Asia,* December 12, 2013, BBC News, http://www.bbc.com/news/science-environment-25210742.

[105] Michael Belfiore, *The Rocketeer,* December 12, 2013, Foreign Policy, http://foreignpolicy.com/2013/12/09/the-rocketeer/.

[106] Elon Musk-Senate Testimony, May 5, 2004.

[107] *Elon Musk: I'll Put a Man on Mars in 10 Years,* April 4, 2011, Market Watch, New York, The Wall Street Journal, http://www.wsj.com/video/elon-musk-ill-put-a-man-on-mars-in-10-years/CCF1FC62-BB0D-4561-938C-DF0DEFAD15BA.html Gwynne Shotwell, March 3, 2014, Broadcast 2212: Special Edition, Interview with Gwynne Shotwell, The Space Show.

[108] https://www.texasmonthly.com/news-politics/elon-musk-boca-chica-starbase-texas/.

[109] Eric Berger, Why Obama's 'giant leap to Mars' is more of a bunny hop right now, October 12, 2016, http://arstechnica.com/science/2016/10/why-obamas-giant-leap-to-mars-is-more-of-a-bunny-hop-right-now/.

[110] Chris Bergin, *SpaceX and Orbital Win Huge CRS Contract from NASA,* December 23, 2008, NASASpaceFlight.com, https://www.nasaspaceflight.com/2008/12/spacex-and-orbital-win-huge-crs-contract-from-nasa/.

[111] *Antares Medium-class Launch Vehicle: Fact Sheet,* Orbital Sciences Corporation PDF, June 3, 2013.

[112] Frank Morring, Jr., Antares Upgrade Will Use RD-181s In Direct Buy From Energomash, December 16, 2014, Aviation Week, http://aviationweek.com/space/antares-upgrade-will-use-rd-181s-direct-buy-energomash.

[113] *Orbital Sciences Orders RD-181 Engines for Antares Rocket,* December 17, 2014, Space News, http://spacenews.com/orbital-sciences-orders-rd-181-engines-for-antares-rocket/.

[114] Boeing and SpaceX Selected to Build America's New Crew Space Transportation System, September 16, 2014, NASA Website, https://blogs.nasa.gov/commercialcrew/2014/09/16/boeing-and-spacex-selected-to-build-americas-new-crew-space-transportation-system/.

[115] Alan Lindenmoyer, *Commercial Crew and Cargo Program,* Annual FAA Commercial Space Transportation Conference

PDF, February 10-11, 2010.

[116] Memi, Edmund G; Gold, Michael N., NASA Selects Boeing for American Recovery and Reinvestment Act Award to Study Crew Capsule-based Design, February 2, 2010, Boeing Press Release, http://boeing.mediaroom.com/2010-02-02-NASA-Selects-Boeing-for-American-Recovery-and-Reinvestment-Act-Award-to-Study-Crew-Capsule-based-Design.

[117] Frank Morring, Jr., Five Vehicles Vie for Future of U.S. Human Spaceflight, April 25, 2011, Aviation Week, http://aviationweek.com/awin/five-vehicles-vie-future-us-human-spaceflight.

[118] https://www.fool.com/investing/2020/07/20/what-went-wrong-with-boeings-starliner-crew-capsul.aspx.

[119] Dreamchaser: Everything You Need to Know About the Mini-Shuttle, April 17, 2014, Human Exploration, http://www.armaghplanet.com/blog/dreamchaser-everything-you-need-to-know-about-the-mini-shuttle.html.

[120] Dave Klinger, Dream Chaser's Crazy Cold War Backstory: The Reusable Mini-spaceplane is Back from the Dead—Again Prepping for Space, September 6, 2012, Infojustice, http://infojustice.org/download/tpp/tpp-news/Leaked_%20US%20proposal%20on%20copyright's%20limits%20_%20Ars%20Technica.pdf.

[121] Adam Higginbotham, *Robert Bigelow Plans a Real Estate Empire in Space,* May 2, 2013, BloombergBusiness, http://www.bloomberg.com/news/articles/2013-05-02/robert-bigelow-plans-a-real-estate-empire-in-space.

[122] Paul Marks, *NASA Turned on by Blow-up Space Stations,* March 3, 2010, New Scientist, https://www.newscientist.com/article/dn18607-nasa-turned-on-by-blow-up-space-stations/.

[123] Ibid.

[124] B*A-2100 Module and other Bigelow Aerospace News,* November 3, 2010, BextBigFuture, http://nextbigfuture.com/2010/10/ba-2100-module-and-other-bigelow.html.

[125] https://www.nzherald.co.nz/business/rocket-lab-unveils-worlds-first-battery-rocket-engine/OODO5DFNZDQZF4ZGQTGAWLVC6I/

[126] https://www.popsci.com/rocket-labs-got-3d-printed-battery-powered-rocket-engine/.

[127] https://spacenews.com/rocket-lab-to-go-public-through-spac-merger-and-develop-medium-lift-rocket/.

[128] Clay Dillow, Blue Origin Makes History By Landing Reusable Rocket A Second Time, January 25, 2019, Fortune.com, http://fortune.com/2016/01/25/blue-origin-makes-history-by-landing-reusable-rocket-a-second-time/.

[129] Elizabeth Howell, *Jeff Bezos: Biography of Blue Origin, Amazon Founder,* January 18, 2013, Space,com, http://www.space.com/19341-jeff-bezos.html.

[130] Calla Cofield, Blue Origin Makes Historic Reusable Rocket Landing in Epic Test Flight, November 24, 2015, Space.com, http://www.space.com/31202-blue-origin-historic-private-rocket-landing.html.

[131] Irene Klotz, *SpaceShipTwo's Rocket Engine Did Not Cause Fatal Crash,* November 3, 2014, Discovery News, http://news.discovery.com/space/private-spaceflight/spaceshiptwos-rocket-engine-did-not-cause-fatal-crash-141103.htm.

[132] Elizabeth Howell, *SpaceShipTwo: On a Flight Path to Space Tourism,* Space.com, February 17, 2016, http://www.space.com/19021-spaceshiptwo.html.

[133] Elizabeth Howell, *SpaceShipTwo: On a Flight Path to Space Tourism,* Space.com, February 17, 2016, http://www.space.com/19021-spaceshiptwo.html.

[134] https://www.space.com/virgin-galactic-richard-branson-unity-22-launch-explained.

[135] https://spaceflightnow.com/2021/04/29/assembly-of-chinese-space-station-begins-with-successful-core-module-launch/.

[136] Cheng Yingqi, New Study Shows Mammals can be Developed in Space, April 17, 2016, Chinadaily.com, http://www.chinadaily.com.cn/china/2016-04/17/content_24611016.htm.

[137] Michael Martina, China Official says Film 'The Martian' Shows Americans want Space Cooperation, April 22, 2016, Reuters, http://www.reuters.com/article/us-china-space-idUSKCN0XJ1C2.

[138] https://arstechnica.com/science/2021/06/rocket-report-china-to-copy-spacexs-super-heavy-vulcan-slips-to-2022/.

[139] https://www.space.com/40114-chandrayaan-1.html.

[140] Ibid.

[141] https://www.space.com/40136-chandrayaan-2.html.

[142] https://www.hindustantimes.com/india-news/chandrayaan3-launch-delayed-further-to-2022-says-isro-chief-k-sivan-101613901105054.html.

[143]https://timesofindia.indiatimes.com/india/after-mars-venus-on-isros-planetary-travel-list/articleshow/69381185.cms.

[144] Pallab Ghosh, *Is Armstrong's Dream Still Alive?*, BBC, 21 July 2009.

[145] https://www.bbc.com/news/science-environment-55998848.

[146] https://www.latimes.com/science/story/2020-09-29/united-arab-emirates-launch-spacecraft-moon.

[147] https://solarsystem.nasa.gov/missions/beresheet/in-depth/.

[148]https://web.archive.org/web/20131224101258/http://www.isa.ir/components1.php?rQV==wHQyAkOklUZnFWdn5WYMJXZ0VWbhJXYw9lZ8B0N3QDQ6QWStVGdp9lZ8BUM4ATMApDZJ52bpR3Yh9lZ.

[149] https://www.esa.int/Science_Exploration/Space_Science/BepiColombo/BepiColombo_factsheet.

[150] Matthew Bodner, *Russia to Propose BRICS Space Station,* January 27, 2015, The Moscow Times, https://themoscowtimes.com/articles/russia-to-propose-brics-space-station-43279.

[151] Edgar Y. Choueiri, *A Critical History of Electric Propulsion: The First 50 Years (1906-1956),* Journal of Propulsion and Power, Vol. 20, No. 2, March-April 2004 http://alfven.princeton.edu/papers/choueiriJPP04a.pdf.

[152] Mark White, *Ion Propulsion 50 Years in the Making,* April 6, 1999, Science News,NASA, http://science.nasa.gov/science-news/science-at-nasa/1999/prop06apr99_2/.

[153] *Robert H. Goddard: American Rocket Pioneer,* Smithsonian Institution Archives, http://siarchives.si.edu/history/exhibits/stories/march-1920-report-concerning-further-developments-space-travel.

[154] Edgar Y. Choueiri, *New Dawn of Electric Rocket: The Ion Drive,* Space Technology, 2009, http://alfven.princeton.edu/papers/sciam2009.pdf.

[155] Michael Cole, *Solar Electric Propulsion: NASA'S Engine to Mars and Beyond",* February 26, 2016, Spaceflight Insider.com, http://www.spaceflightinsider.com/missions/human-spaceflight/solar-electric-propulsion-nasas-engine-mars-beyond/.

[156] Ibid.

[157] Ibid.

[158] Myers, Roger and Carpenter, Christian, High Power Solar Electric Propulsion for Human Space Exploration Architectures, IEPC 2011-261, September 11-15, 2011.

[159] Stephen Clark, NASA Eyes Ion Engines for Mars Orbiter Launching in 2022, March 3, 2015.

[160] Lee Billings, A Rocket for the 21st Century: Former Astronaut Franklin Chang-Diaz Explains How His Plasma Rocket Engine Could Revolutionize Space Travel and Why We Need Nuclear Power in Space, September 29, 2009, Seed Magazine.com, http://seedmagazine.com/content/print/a_rocket_for_the_21st_century/.

[161] Beth Dickey, Star Power: The Plasma Rocket, Says U.S. Astronaut Franklin Chang-Diaz, is the Propulsion Technology of the Future, March 2004, Air & Space.com, http://www.airspacemag.com/space/star-power-6700758/#AOW2dfMFrWGxsc25.99.

[162] Lee Billings, A Rocket for the 21st Century: Former Astronaut Franklin Chang-Diaz Explains How His Plasma Rocket Engine Could Revolutionize Space Travel and Why We Need Nuclear Power in Space, September 29, 2009, Seed Magazine.com, http://seedmagazine.com/content/print/a_rocket_for_the_21st_century/.

[163] Squire, Jared P, et.al, *High Power VASIMR Experiments using Deuterium, Neon and Argon*, International Electric Propulsion Conference, 2007 http://www.adastrarocket.com/Jared_IEPC07.pdf.

[164] *VASIMR VX-200 First Stage Achieves Full Power Raring*, October 24, 2008, Ad Astra Rocket Company Press Release, http://www.adastrarocket.com/Release241008.pdf.

[165] NASA Announces New Partnerships with U.S. Industry for Key Deep-Space Capabilities, March 30, 2015, NASA Press Release, http://www.nasa.gov/press/2015/march/nasa-announces-new-partnerships-with-us-industry-for-key-deep-space-capabilities.

[166] Lee Billings, A Rocket for the 21st Century: Former Astronaut Franklin Chang -Diaz Explains How His Plasma Rocket Engine Could Revolutionize Space Travel and Why We Need Nuclear Power in Space, September 29, 2009, Seed Magazine.com, http://seedmagazine.com/content/print/a_rocket_for_the_21st_century/.

[167] *Nuclear Reactors and Radioisotopes for Space,* World Nuclear Association, Updated February, 2016,

http://www.world-nuclear.org/information-library/non-power-nuclear-applications/transport/nuclear-reactors-for-space.aspx.

[168] *Nuclear Reactors and Radioisotopes for Space,* World Nuclear Association, Updated February, 2016, http://www.world-nuclear.org/information-library/non-power-nuclear-applications/transport/nuclear-reactors-for-space.aspx.

[169] *The SP-100 Nuclear Reactor Program: Should it Be Continued?,* U.S. Government Accountability Office, T-NSIAD-92-15, March 1992, http://www.gao.gov/products/T-NSIAD-92-15.

[170] *Thorium,* World Nuclear Association Library, http://www.world-nuclear.org/information-library/current-and-future-generation/thorium.aspx.

[171] Cindy Hurst, *Fuel for Thought: The Importance of Thorium to China,* February 2015, Institute for the Analysis of Global Security (IAGS), http://iags.org/thoriumchina.pdf.

[172] Ambrose Evans-Pritchard, *China Blazes Trail for 'Clean' Nuclear Power from Thorium,* January 6, 2013, Telegraph, http://www.telegraph.co.uk/finance/comment/ambroseevans_pritchard/9784044/China-blazes-trail-for-clean-nuclear-power-from-thorium.html.

[173] Sebastian Anthony, Thorium Nuclear Reactor Trial Begins, Could Provide Cleaner, Safer, Almost-waste-free Energy, July 1, 2013, Extreme Tech, http://www.extremetech.com/extreme/160131-thorium-nuclear-reactor-trial-begins-could-provide-cleaner-safer-almost-waste-free-energy.

[174] *Thorium,* World Nuclear Association Library, http://www.world-nuclear.org/information-library/current-and-future-generation/thorium.aspx.

[175] Dino Grandoni, *Why It's Taking the U.S. So Long to Make Fusion Energy Work,* January 26, 2026, HuffPost Tech, http://www.huffingtonpost.com/2015/01/20/fusion-energy-reactor_n_6438772.html.

[176] John Hewitt, *China is Going to Mine the Moon for Helium-3 Fusion Fuel,* January 26, 2015 Extreme Tech, http://www.extremetech.com/extreme/197784-china-is-going-to-mine-the-moon-for-helium-3-fusion-fuel.

[177] Ibid.

[178] Ameera David, *The Lunar Reconnaissance Orbiter & Water on the Moon,* January 9, 2015, The Erimtan Angle, RT, https://sitanbul.wordpress.com/2015/01/10/the-lunar-reconnaissance-orbiter-water-on-the-moon/.

[179] Yvonne J. Pendleton, *Water on the Moon,* Astronomy in Focus, Volume 1, XXIXth IAU General Assembly, 2015 International Astronomical Union.

[180] Paul Spudis, *Lunar Resources: Unlocking the Space Frontier,* Ad Astra, Volume 23 Number 2, Summer 2011, National Space Society, http://www.nss.org/adastra/volume23/lunarresources.html.

[181] Mike Wall, NASA is Studying How to Mine the Moon for Water, October 9, 2014, Space.com, http://www.space.com/27388-nasa-moon-mining-missions-water.html.

[182] Mining the Moon for Lunar Resources, PERMANENT.com, http://permanent.com/mining-the-moon-for-lunar-resources.html

[183] Jean-Louis Santini, *NASA Bets on Private Companies to Exploit Moon's Resources,* February 9, 2014, Phys Org, Astronomy & Space/Space Exploration, http://phys.org/news/2014-02-nasa-private-companies-exploit-moon.html.

[184] Joshua Philipp, Mining the Moon: Plans Taking Off, but Rules Lacking, January 19, 2014, Epoch Times, http://www.theepochtimes.com/n3/476806-mining-the-moon-plans-taking-off-but-rules-lacking/.

[185] Ibid.

[186] Alan Boyle, Moon Express asks FAA to Review its Plans for Google Lunar X Prize Landing in 2017, April 8, 2016, Geek Wire.com, http://www.geekwire.com/2016/moon-express-asks-faa-review-payload-lunar-landing-2017/.

[187] https://spacenews.com/google-lunar-x-prize-to-end-without-winner/

[188] https://www.space.com/israeli-beresheet-moon-landing-attempt-fails.html

[189] NASA Confirms Evidence That Liquid Water Flows on Today's Mars, NASA Press Release, September 28, 2015, http://www.nasa.gov/press-release/nasa-confirms-evidence-that-liquid-water-flows-on-today-s-mars.

[190] Ibid.

[191] Larry O'Hanlon, *Mining Mars? Where's the Ore?,* February 22, 2010, Discovery News, http://news.discovery.com/space/history-of-space/mars-prospecting-ores-gold.htm.

[192] Ernst, R. 2007, Large Igneous Provinces in Canada through Time and Their Metallogenic Potential. Mineral Deposits of Canada: A Synthesis of Major Depotit-Types, District Metallogeny, the Evolution of Geological Provinces, and Exploration Methods, Geological Association of Canada, Mineral Division, Special Publication No. 5. 929-937.

[193] Hugh H. Kieffer (1992), *Mars,* University of Arizona Press, ISBN 978-0-8165-1257-7.

[194] Ruzicka, G. et al, (2001), Comparative Geochemistry of Basalts from the Moon, Earth, HED Asteroid, and Mars:Iimplications for the Origin of the Moon, Geochimica et Cosmochimica ACTA: 65. 979-997, http://www.sciencedirect.com/science/article/pii/S0016703700005998.

[195] Head, J., et al, 2006, The Huygens-Hellas Giant Dike System on Mars: Implications for Late Noachian-Early Hesperian Volcanic Resurfacing and Climate Evolution, Geology: 34. 285-288, http://www.planetary.brown.edu/pdfs/3256.pdf.

[196] Calla Cofield, *Search for 'Missing' Carbon on Mars Cancelled,* November 26, 2015, Space.com, http://www.space.com/31215-mars-missing-carbon-mystery.html.

[197] Francois Forget, *Alien Weather at the Poles of Mars,* November 19, 2004, Science.org, http://science.sciencemag.org/content/306/5700/1298.

[198] NASA Mars Lander Sees Falling Snow, Soil Data Suggests Liquid Past, September 29, 2008, NASA Website, http://www.nasa.gov/mission_pages/phoenix/news/phoenix-20080929.html.

[199] *Mars Clouds Higher Than Any On Earth,* August 28, 2006, Space.com, http://www.space.com/2812-mars-clouds-higher-earth.html.

[200] *Mars Facts*, NASA Website, http://mars.nasa.gov/allaboutmars/facts/.

[201] James E. Tillman, *Mars-Temperature Overview,* University of Washington, http://wwwk12.atmos.washington.edu/k12/resources/mars_data_information/temperature_overview.html.

[202] *Extreme Planet Takes its Toll,* June 12, 2007, NASA Website, Feature, http://www.nasa.gov/mission_pages/mer/mars_mer_feature_20070612.html.

[203] Nola Taylor Red, *Mars' Moons: Facts About Phobos & Deimos,* March 27, 2013, Space.com, http://www.space.com/20413-phobos-deimos-mars-moons.html.

[204] Sputnik News, RIA Novosti, *China Will Send a Space Probe to Mars with Russia's Assistance in October,* December 5, 2008, http://sputniknews.com/world/20081205/118704675.html#ixzz47AQjjBmP.

[205] Yuri Zaitsev, *Russia to Study Martian Moons Once Again,* July 14, 2008, Sputnik News, RIA Novosti, http://sputniknews.com/analysis/20080714/113951848.html.

[206] Buzz Aldrin and Leonard David, *Mission to Mars: My Vision for Space Exploration*, National Geographic Books, 2013 (p. 17).

[207] Ibid. (p. 87).

[208] Ibid. (p. 27).

[209] Ibid. (p. 26).

[210] Ibid. (p. 6).

[211] Ibid. (p. 8).

[212] Ibid. (p. 8).

[213] Ibid. (p. 9).

[214] Ibid. (p. 9).

[215] Ibid. (p. 13).

[216] Ibid. (p. 13).

[217] Ibid. (p. 14).

[218] Ibid. (p. 15).

[219] Ibid. (p. 15).

[220] Ibid. (p. 13).

[221] Ibid. (p. 13).

[222] Ibid. (p. 14).

[223] Ibid. (p. 17).

[224] Ibid. (p. 85).

[225] Ibid. (p. 87).

[226] Ibid. (p. 88).

[227] Ibid. (p. 38-39).

[228] Ibid. (p. 46).

[229] Ibid. (p. 55).

[230] Ibid. (p. 56).

[231] Ibid. (p. 59).

[232] Ibid. (p. 60).

[233] Ibid. (p. 201).

[234] Ibid. (p. 201-202).

[235] Ibid. (p. 202).

[236] Ibid. (p. 18).

[237] Ibid. (p. 18).

[238] Ibid. (p. 104-105).

[239] Ibid. (p. 105).

[240] Ibid. (p. 105).

[241] Ibid. (p. 106).

[242] Ibid. (p. 106).

[243] Ibid. (p. 106-107).

[244] Ibid. (p. 107).

[245] Ibid. (p. 108).

[246] Ibid. (p. 100-102).

[247] Ibid. (p. 104).

[248] Ibid. (p. 165-166).

[249] Ibid. (p. 172-173).

[250] Ibid. (p. 165-166).

[251] Ibid. (p. 166).

[252] Ibid. (p. 167).

[253] Ibid. (p. 171).

[254] Ibid. (p. 171).

[255] Ibid. (p. 169-170).

[256] Ibid. (p. 171).

[257] Ibid. (p. 42).

[258] Ibid. (p. 45).

[259] Ibid. (p. 98).

[260] Ibid. (p. 98).

[261] Ibid. (p. 98).

[262] Ibid. (p. 100).

[263] Ibid. (p. 101).

[264] Ibid. (p. 146).

[265] Ibid. (p. 147).

[266] Ibid. (p. 148).

[267] Ibid. (p. 145-146).

[268] Ibid. (p. 149-150).

[269] Ibid. (p. 151).

[270] Ibid. (p. 161-162).

[271] Ibid. (p. 158).
[272] Ibid. (p. 158).
[273] Ibid. (p. 178).
[274] Ibid. (p. 173).
[275] Ibid. (p. 181).
[276] Ibid. (p. 177).
[277] Ibid. (p. 183).
[278] Ibid. (p. 186-187).
[279] Ibid. (p. 195).
[280] Ibid. (p. 195).
[281] Ibid. (p. 195).
[282] Ibid. (p. 196).
[283] Ibid. (p. 196).
[284] Ibid. (p. 196).
[285] Ibid. (p. 196).
[286] Ibid. (p. 197).
[287] Ibid. (p. 197).
[288] Ibid. (p. 197).
[289] Ibid. (p.198).
[290] Ibid. (p. 198-199).
[291] Ibid. (p. 199).
[292] Ibid. (p. 199).
[293] Ibid. (p. 199).
[294] Ibid. (p. 117).
[295] Ibid. (p. 118).
[296] Ibid. (p. 120).
[297] Ibid. (p. 121).
[298] Ibid. (p. 136).
[299] Ibid. (p. 202).
[300] Ibid. (p. 202-203).
[301] Ibid. (p. 203).
[302] Ibid. (p. 204).
[303] Ibid. (p. 204).
[304] Ibid. (p. 204-205).
[305] Ibid. (p. 205).
[306] Ibid. (p. 205).
[307] Ibid. (p. 205-206).
[308] Ibid. (p. 207).
[309] Ibid. (p. 207).
[310] Ibid. (p. 207).
[311] Ibid. (p. 207).
[312] Ibid. (p. 209).
[313] Ibid. (p. 209).
[314] Ibid. (p. 209).

www.ingramcontent.com/pod-product-compliance
Lightning Source LLC
Chambersburg PA
CBHW061325190326

41458CB00011B/3901